THE McGRAW-HILL CIVIL ENGINEERING PE EXAM DEPTH GUIDE

Water Resources

THE McGRAW-HILL CIVIL ENGINEERING PE EXAM DEPTH GUIDE

Water Resources

Emmanuel U. Nzewi, Ph.D., PE

McGRAW-HILL

New York Chicago San Francisco Lisbon London Madrid
Mexico City Milan New Delhi San Juan Seoul
Singapore Sydney Toronto

Cataloging-in-Publication Data is on file with the Library of Congress

McGraw-Hill

A Division of The McGraw-Hill Companies

1 2 3 4 5 6 7 8 9 0 AGM/AGM 0 7 6 5 4 3 2 1

ISBN 0-07-136183-9

The sponsoring editor for this book was Larry S. Hager and the production supervisor was Sherri Souffrance. It was set in Times Roman by Lone Wolf Enterprises, Ltd.

Printed and bound by Quebecor/Martinsburg.

This book is printed on recycled, acid-free paper containing a minimum of 50% recycled, de-inked fiber.

CONTENTS

PREFACE

The main focus of this book is to provide the reader with a concise review of topics in water resources engineering (hydraulics and hydrology). The intended reader is an individual preparing to sit for the principles and practice exam (often called the PE exam) conducted by the National Council of Examiners for Engineering and Surveying (NCEES). The book is divided into six sections—five chapters and an appendix. The entire book is illustrated with diagrams and figures which help provide the reader with a more concrete description of the engineering concepts discussed. In general, each major concept presented is followed by an illustrative and non-trivial example. It is hoped that these features of the book will make the review of water resources engineering topics more efficient because examples normally help to answer questions that the reader may have, and diagrams convey ideas more effectively than mere words.

Chapters One and Two provide an overview of the fundamental principles necessary for a good understanding of the topics presented in Chapter Three. Thus, Chapters One and Two are the foundation of the rest of the book. While some topics covered in Chapter One may be somewhat elementary, it is the belief of the author that incorrect assumptions usually spring from a poor understanding of these fundamental concepts. Chapter Two includes detailed presentations of fluid statics, basic principles in manometry and flotation of objects.

In Chapter Three, hydraulics are presented along with illustrative examples. The reader is encouraged to work out the example problems and then walk through the provided solutions for correction, reinforcement, and reflection.

Chapter Four is dedicated to engineering hydrology. This part of the book presents an overview of engineering hydrology and culminates in the implementation approaches for the Rational or Lloyd-Davies formula, Unit Hydrograph techniques, Natural Resources Conservation Service (NRCS) methods (especially the Curve Number method), and their application in storm water management.

In Chapter Five, a survey and overview of the major concepts in groundwater hydrology. Topics include classification of groundwater, types of aquifers, basic parameters in groundwater storage and movement, Darcy's Law, and the Laplace equation for groundwater flow and flow nets. Example problems are provided to demonstrate the proper application of these groundwater hydraulics concepts in solving engineering problems.

I would like to acknowledge the support of my wife, Sochi Caroline, and kids—Sammy, 'Gechi, Joshy, Ikenna and Toby. The input of my students (in regular classes and review courses) on the technical aspects of the development of this book has been invaluable. The assistance of the individuals who prepared some of the manuscript (Cheryl Simmons and Sherry Bittle) and others who supported me in the completion of this project is hereby noted and greatly appreciated. The persistence and encouragement of the *Lone Wolf* crew in bringing this book to fruition is also acknowledged. Above all, I thank my God, who provided the strength, grace, and perseverance to complete this work. To Him be all the glory.

Emmanuel U. Nzewi, Ph.D., PE

ABOUT THE AUTHOR

EMMANUEL U. NZEWI, Ph.D., PE

Emmanuel U. Nzewi, Ph.D., PE, has expert credentials in a number of water resources areas, including hydropower production, urban watershed analysis, modeling, and water distribution. A professor of civil engineering at North Carolina A&T, he earned his doctorate from Purdue University.

CHAPTER 1
FUNDAMENTAL CONCEPTS AND PROPERTIES OF FLUIDS

In this chapter, the properties of fluids (gases and liquids) that are of interest to the civil engineer are presented. It is assumed that the fluid is a continuum. The fluid properties described include: density, specific weight, specific gravity, viscosity and the classification of fluids as Newtonian and non-Newtonian, the no-slip condition, surface tension, compressibility of fluids and vapor pressure concepts. The specific capacity of gases is briefly discussed. At the beginning of the chapter, a set of basic equations in hydraulics are listed and summarily described. In general, Chapter One covers foundational issues with reference to fluid properties. Although fluids are generally discussed, the emphasis is on liquids, particularly water.

1.0 OVERVIEW AND LIST OF REFERENCE FORMULAS

This chapter presents the principles of basic fluid mechanics necessary to solve problems in hydraulics and hydrostatics. Because of the review nature of this text, these topics are summarily presented with illustrative examples where applicable. Pertinent tables and figures are provided liberally so that the reader does not need to consult other reference texts or publications for common reference values and formulas.

1.01 List of Formulas

1. Density traditionally is represented by the Greek letter ρ.

$$\rho = \frac{\text{Mass}}{\text{Volume}} = \frac{\text{M}}{\forall}$$

2. Specific weight traditionally is represented by the Greek letter γ.

$$\gamma = \frac{\text{Weight}}{\text{Volume}} = \frac{\text{W}}{\forall}$$

3. Specific gravity (s)

$$s = \frac{\text{Density of substance}}{\text{Density of water@ 4°C}} = \frac{\rho}{\rho \text{ of water@4°C}}$$
$$= \frac{\text{Specific weight of substance}}{\text{Specific weight of water@4°C}}$$
$$= \frac{\gamma}{\gamma \text{ of water@4°C}}$$

4. Absolute viscosity traditionally is represented by the Greek letter μ.

$$\mu = \frac{\text{Shear Stress}}{\text{Strain Rate}} = \frac{\tau}{\frac{du}{dy}}$$

therefore, shear stress, $\tau\left[\frac{N}{m^2}\right] = \mu\,\frac{du}{dy}$

5. Kinematic viscosity (ν)

$$\nu = \frac{\text{Absolute Viscosity}}{\text{Density}} = \frac{\mu}{\rho}$$

6. Surface tension traditionally is represented by the Greek letter σ.

$$\sigma = \frac{h\rho D}{4\cos\beta}$$

where β = Angle of contact
h = Capillary rise
D = Diameter of capillary tube
ρ = Density of liquid.

When β is acute (<90°) capillary rise results (e.g., water/glass).

When β is obtuse (>90°) capillary depression results (e.g., mercury/glass).

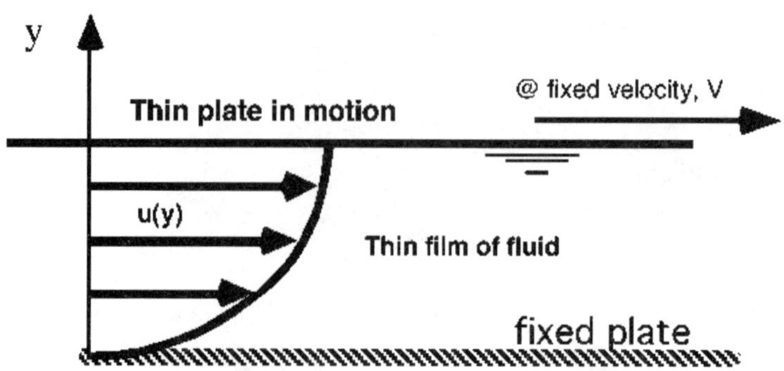

FIGURE 1.0.1

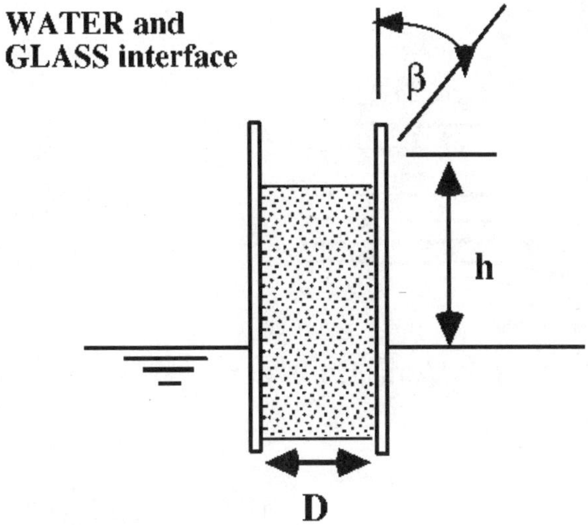

FIGURE 1.0.2 Water and glass.

7. Pressure at a point

$$\text{Pressure} = \frac{\text{Normal Force}}{\text{Normal Area}} = \frac{F}{A}$$

Pressure is a scalar quantity (with a magnitude but no direction)

8. Pressure at a point in a fluid

$$\begin{aligned}\text{Pressure} &= (\text{specific weight of fluid})(\text{depth in fluid}) \\ &= \gamma h\end{aligned}$$

9. Force (F) on a submerged surface in fluid

$$F = \gamma \times \bar{h} \times \text{area}$$

$$\bar{h} = \text{depth to the centroid of the surface (with area} = A)$$

Therefore: $F = \gamma \times \bar{h} \times A$

10. Archimedes' Principle—for a fully or partially submerged object

$$\text{Buoyant force} = \gamma \forall$$

where $\forall$ = Volume of fluid displaced by the object
γ = Specific weight of the fluid.

11. Bernoulli equation (Caution: not the same as the energy equation)

Assuming an inviscid (viscous-less) fluid, that flow is incompressible and for two points [1] and [2] on a streamline:

$$\frac{P}{\gamma} + z + \frac{V^2}{2g} = \text{constant}$$

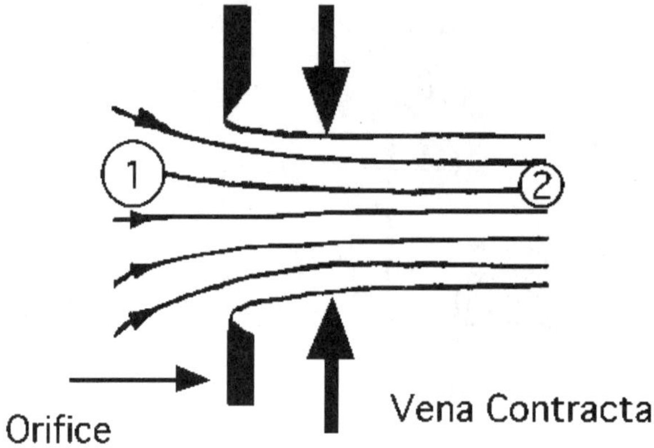

FIGURE 1.0.3 Flow along a streamline.

which is:

$$\frac{P_1}{\gamma} + z_1 + \frac{V_1^2}{2g} = \frac{P_2}{\gamma} + z_2 + \frac{V_2^2}{2g}$$

which can be summarily described as:

(pressured head) + (elevation head) + (velocity head) = constant

12. Continuity equation (mass conservation) for two sections in a stream tube (or pipe)

$$\rho_1 V_1 A_1 = \rho_2 V_2 A_2$$

where ρ_1, ρ_2 = Density of fluid at sections 1 and 2
V_1, V_2 = Velocity of sections 1 and 2
A_1, A_2 = Cross-sectional area at sections 1 and 2.

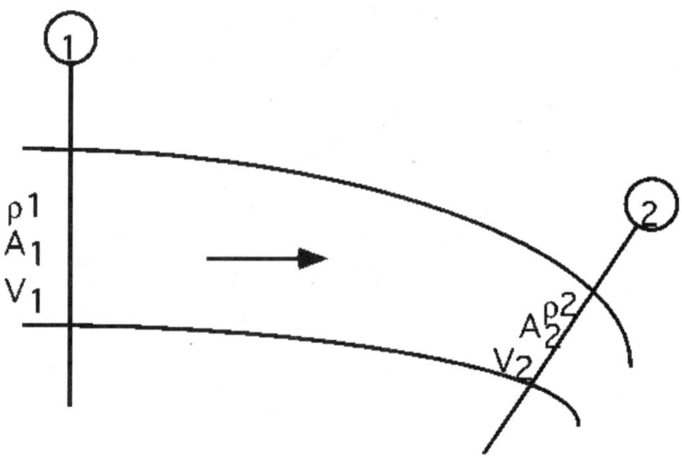

FIGURE 1.0.4 Flow through a steam tube.

13. Energy equation (Energy can change form but remains constant)

$$\frac{P_1}{\gamma} + \frac{\alpha_1 V_1^2}{2g} + z_1 + h_p = \frac{P_2}{\gamma} + \frac{\alpha_1 V_2^2}{2g} + z_2 + \Sigma h_f + \Sigma h_m + h_t$$

where P_1, P_2 = Pressure in flow at sections 1 and 2
V_1, V_2 = Velocity in flow at sections 1 and 2
Z_1, Z_2 = Positional elevation at sections 1 and 2
α_1, α_2 = Kinetic energy flux correction factors at 1 and 2
h_p = Head supplied by a pump
Σh_p = Total friction head loss between 1 and 2
Σh_m = Total minor head losses between 1 and 2
Σh_t = Head extracted by a turbine or heat exchanger.

14. Friction head loss using Darcy-Weisbach equation

$$h_f = f \frac{L}{D} \frac{V^2}{2g}$$

where f = Friction factor (obtained from Moody Diagram)
L = Length of pipe
D = Diameter of pipe
V = Velocity of flow
g = Gravitational acceleration.

15. Minor head loss—due to fittings, values, and changes in flow area/shape/direction

$$h_m = K \frac{V^2}{2g}$$

where K is the minor head loss coefficient, which can be obtained from appropriate standard reference tables.

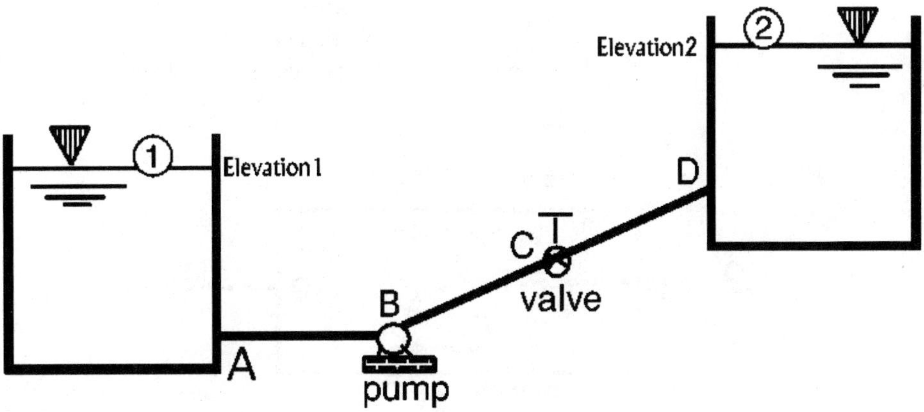

FIGURE 1.0.5 Flow through a typical piping system.

16. Simple pipe network (branched system)

a. Head loss across [A]–[B]is the same regardless of path (through [1]or [2])

$$\text{Therefore, } \frac{f_1 L_1 V_1^2}{2gD_1} = \frac{f_2 L_2 V_2^2}{2gD_2} \Rightarrow \frac{f_1 L_1 V_1^2}{2D_1} = \frac{f_2 L_2 V_2^2}{2D_2}$$

where f, V, and g are the friction factor, velocity of flow and gravitational acceleration, respectively.

b. The *continuity condition* implies that the algebraic sum of flows at a junction must equal zero.

$$\text{Therefore, } Q = Q_1 + Q_2$$

and thus, for circular pipes:

$$Q = \frac{1}{4}\pi(D_1^2 V_1 + D_2^2 V_2)$$

17. Hazen-Williams Head Loss Equation

$$h_f = 3 \cdot 02 \frac{L}{D^{1 \cdot 167}}\left(\frac{V}{C}\right)^{1 \cdot 85}$$

where D = Diameter of pipe
V = Velocity of flow
L = Length of pipe
C = Hazen-Williams C.

18. Manning's Equation—considering open channel flow

$$V = \frac{k}{n} R^{2/3} \sqrt{S}$$

where k = 1.0 (SI units) = 1.486 (English units)
n = Manning's roughness factor for the channel
R = Hydraulic radius = $\dfrac{\text{Area of flow}}{\text{Wetted perimeter}}$
S = Longitudinal slope of open channel or slope of energy grade line.

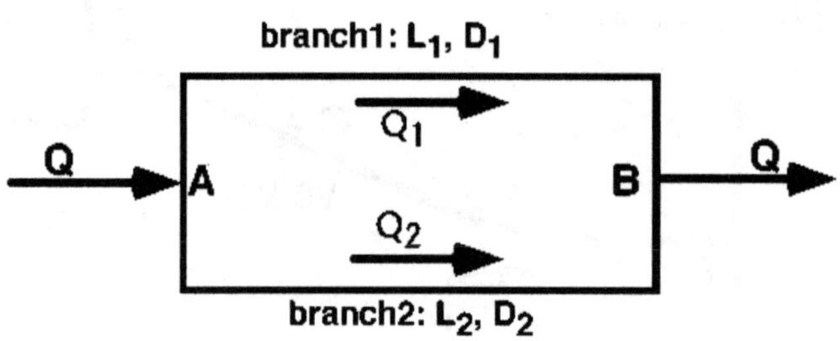

FIGURE 1.0.6 Parallel pipe system.

19. Reynolds Number—ratio of viscous to inertial forces in a flow

$$\text{Re} = \frac{VD}{\nu}$$

where Re < 2000 = Laminar flow
Re > 3000 = Turbulent flow (practical limit is 4000)
V = Velocity
D = Characteristic length (for pipes = diameter)
ν = Kinematic viscosity of the fluid flowing.

20. Friction factor equations

Swamee and Jain equation:

a. $$f = 1.325\left[\ln\left(0.27\frac{\epsilon}{D} + \frac{5.74}{\text{Re}^{0.9}}\right)\right]^{-2}$$

when $5 \times 10^3 \leq \text{Re} \leq 10^8$ and $10^{-6} \leq \frac{\epsilon}{D} \leq 10^{-2}$

where ϵ = Equivalent sand roughness size of pipe material
D = Diameter of pipe (in the same units as ϵ)
Re = Reynolds number.

b. When the flow is in fully rough turbulent zone (fully rough regime is to the right of the transitional zone in the Moody Diagram)

$$f = 1.325\left[\ln\left(0.27\frac{\epsilon}{D}\right)\right]^{-2}$$

21. Application of Swamee and Jain equation:

a. To calculate flow rate (Q) in a pipe when the head loss, h_f is given

$$Q = -0.965D^2\left[\sqrt{\frac{gDh_f}{L}}\right]\ln\left[0.27\frac{\epsilon}{D} + \frac{1.784\nu}{\sqrt{\dfrac{gDh_f}{L}}}\right]$$

when Re ≥ 3000 (practically 4000)

b. Calculate pipe diameter (D) for given Q and h_f

$$D = 0.66\left[\epsilon^{1.25}\left(\frac{LQ^2}{gh_f}\right)^{4.75} + \nu Q^{9.4}\left(\frac{L}{gh_f}\right)^{52}\right]^{0.04}$$

when $3000 \leq \text{Re}\ 3 \times 10^8$ and $10^{-6} \leq \frac{\epsilon}{D} \leq 2 \times 10^{-2}$

22. Friction factor when flow is Laminar (Re < 2000)

$$\text{Friction factor}, f = \frac{64}{\text{Re}}$$

23. Flow rate (Q) equations for circular pipes in a variety of units

$$Q\left[\frac{ft^3}{s}\right] = 0.7854D^2V$$

$$Q\left[\frac{ft^3}{s}\right] = 0.005454d^2V$$

$$q[gpm] = 352.488D^2V$$

$$q[gpm] = 2.448d^2V$$

where d = Diameter in inches
D = Diameter in feet
V = Velocity in ft/s
q = Flow rate in gallons per minute (gpm)

24. Pipe diameter for a sewer segment laid at a slope, S for a design flow

$$Q_{design} \geq Q_{peak}$$

$$D\,[ft] = \left[\frac{4^{5/3}Q_{design} \times n}{1.486\pi\sqrt{s}}\right]^{3/8}$$

where n = Manning's roughness coefficient for the pipe material

Choose the next commercially available pipe size to the calculated value D

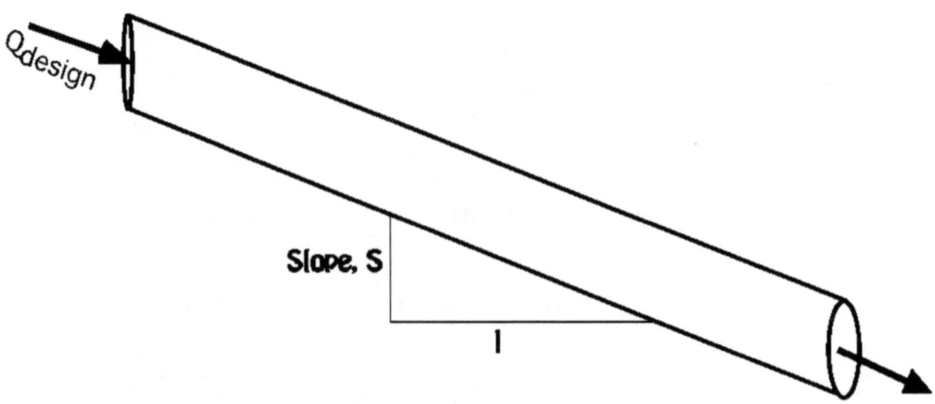

FIGURE 1.0.7 Segment of a sewer system.

25. Momentum Equation for pipe flow

$$\Sigma F_x = m(V_{2x} - V_{1x}) = \rho Q\,(V_{2x} - V_{1x})$$
$$\Sigma F_y = m(V_{2y} - V_{1y}) = \rho Q'\,(V_{2y} - V_{1y})$$
$$\Sigma F_z = m(V_{2z} - V_{1z}) = \rho Q\,(V_{2z} - V_{1z})$$

26. Ideal Gas Law

$$\text{Density, } \rho = \frac{P}{RT}$$

where P = Pressure in absolute units
R = Gas constant
T = Absolute temperature [K or °R]

1.1 FUNDAMENTAL IDEAS AND DEFINITIONS

The term fluid refers to a substance that deforms continuously when subjected to shear stress. Thus, fluids cannot resist shear stresses, in contrast to solids, which return to their original shape (as long as the yield limit is not exceeded) when the deforming stresses are removed. Fluids include liquids and gases.

A liquid has a fixed volume, but no fixed shape. A gas has neither fixed volume nor shape and will expand to fill its container. Examples of liquids include water, alcohol, motor oil, toothpaste, grease, liquid nitrogen, and gasoline. Gases include air (which is a mixture of gases, including water vapor), ozone, carbon dioxide, methane, carbon monoxide, and nitrogen.

Gases are considered to be a continuum unless in situations where gases are at extremely low pressures (rarefied gases). The idea of fluids as a continuum under normal conditions is important when presenting properties calculated for gases as being representative of the *collection of molecules*. Unless a continuum can be assumed, uniform properties cannot be evaluated theoretically and are not justified. Because fluid properties are assumed to vary continuously throughout the liquid or gas, the concept of a continuum is necessary in fluid mechanics.

The remainder of this chapter presents basic fluid mechanics principles that describe and discuss the behavior of fluids (gases and liquids) at rest (hydrostatics).

1.2 PROPERTIES OF FLUIDS

Presented below are selected properties of gases and liquids. Where applicable, tables and graphs are included to provide the reader with reference values and essential facts.

1.2.1 Density (ρ) and Specific Weight (γ)

$$\text{Density, } \rho = \frac{\text{Mass of Substance}}{\text{Volume of substance}} = \frac{\text{M}}{\forall}$$

$$\text{Units} = \left[\frac{\text{kg}}{\text{m}^3} \text{ or } \frac{\text{slugs}}{\text{ft}^3}\right]$$

The density of a liquid decreases as temperature increases. Under constant pressure, the density of gases decreases as well. Pressure changes usually have insignificant impact on the density of liquids.

Specific weight $$\gamma = \frac{\text{Weight of fluid}}{\text{Volume of fluid}} = \frac{Mg}{\forall} = \rho g$$

Specific gravity $$s = \frac{\rho_{fluid}}{\rho_{water\ @4°C}}$$

Anomalous behavior of water. In general, the density of water decreases as the temperature increases (see Figure 1.2.1.1 and Table 1.2.1.1). It is observed that the maximum density of water, however, occurs at 3.98°C. Thus, ice (temperature $\leq$ 0°C) floats on (warmer) water at 3.98°C. This is the reason lakes, in colder climates, freeze from the surface toward the bottom. This property of water helps to preserve aquatic life (fish do not usually freeze in frozen lakes) and serves as a design feature that preserves life on earth in colder regions. See Table 1.2.1.1.

The following equations give the density (ρ) and specific weight (γ) of water within 1 percent of the measured values at different temperatures, T [°C]).

$$\rho = 1000 - \frac{(T-4)^2}{180}$$

$$\gamma = 9810 - \frac{(T-4)^2}{18}$$

EXAMPLE PROBLEM 1.2.1.1

Suppose 20 liters of a given liquid weighs 25.0 N

Determine the density $\left[\frac{\text{kg}}{\text{m}^3}\right]$

Calculate the specific gravity

TABLE 1.2.1.1. Density of Water around 4°C

Temperature [°C]	Density [kg/m3]
–10	998.15
–5	999.30
0	999.87
2	999.97
4(3.98)	1000.00
6	999.97
8	999.88
10	999.73
12	999.53

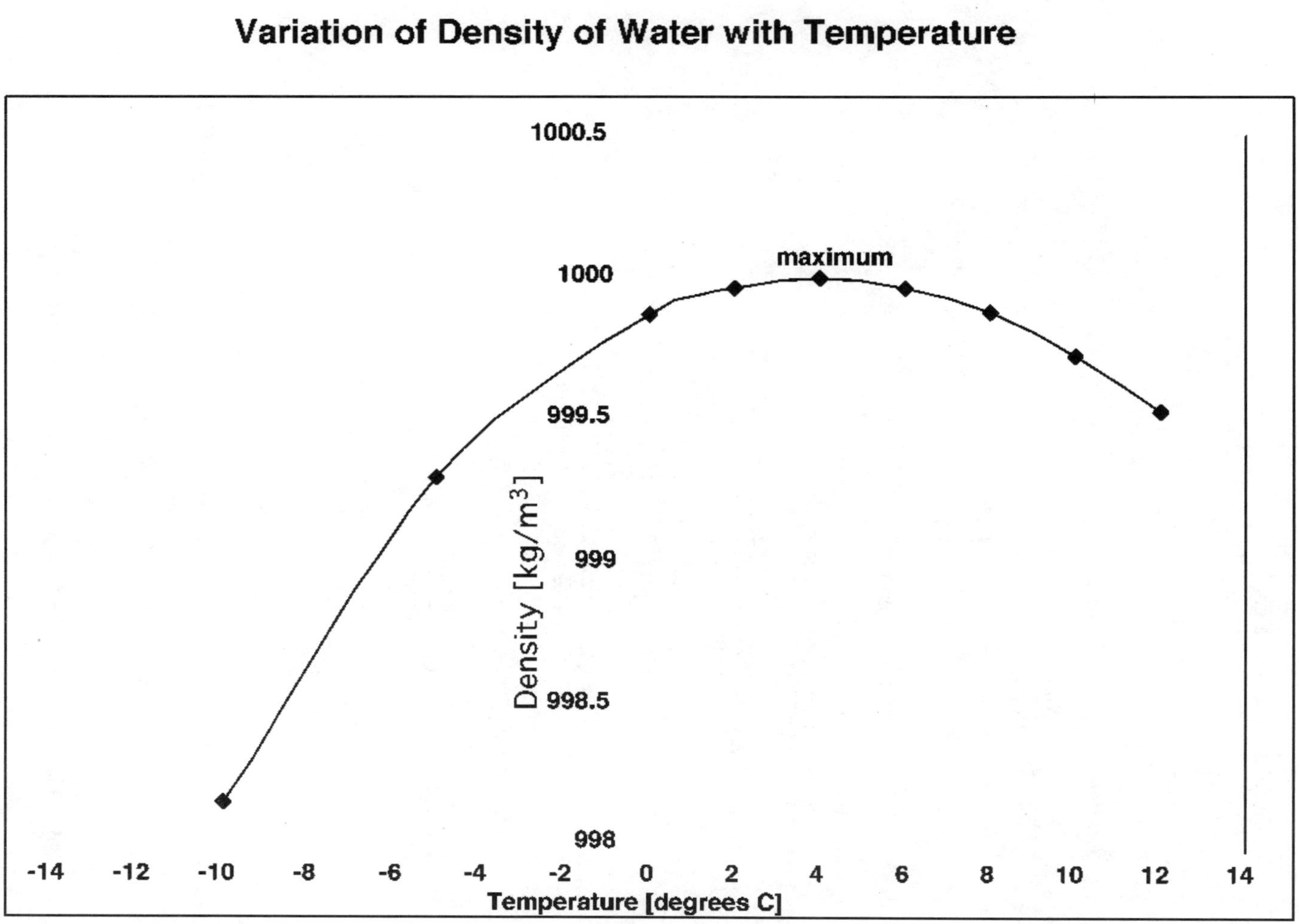

FIGURE 1.2.1.1 Variation of Density of Water with Temperature around 4°C.

Calculate specific weight $\left[\frac{\text{N}}{\text{m}^3}\right]$

Assume that $g = \frac{9.81\text{ m}}{\text{s}^2}$.

Solution:

- Density $= \frac{\text{Mass}}{\text{Volume}} = \rho$

$$(\text{weight}) = \left(\overset{M}{\text{Mass}}\right) \times \left(\text{gravitation}\overset{g}{\ }\text{acceleration}\right)$$

$$\therefore \text{mass} = \frac{\text{weight}}{\text{g}} = \frac{25.0\ N}{9.81\text{ m/s}^2} = 2.548\text{ kg}$$

Note: $1\text{ N} = 1\text{ kg}\frac{\text{m}}{\text{s}^2}$ and $1\text{ liter} = 10^{-3}\text{ m}^3$

But $\frac{\text{m}}{\text{s}^2}\ \rho = \frac{\text{Mass}}{\text{Volume}} = \frac{2.548\text{ kg}}{20 \times 10^{-3}}$

Therefore, $\rho = 127.42\ \frac{\text{kg}}{\text{m}^3}$

- Specific gravity, $\text{s} = \frac{\rho_{\text{fluid}}}{\rho_{\text{water@4°C}}} = \frac{127.42\text{ kg/m}^3}{1000\text{ kg/m}^3} = 0.13$

Note: ρ water @ 4°C $= 1000\ \frac{\text{kg}}{\text{m}^3}$

- Specific weight, $\gamma = \frac{\text{Weight}}{\text{Volume}} = \rho g$

$$= \frac{25N}{20 \times 10^{-3}N} = (127.42)(9.81)$$

$$= 1250\ \frac{N}{\text{m}^3} = 1250\ \frac{N}{\text{m}^3}$$

Density of Gases

Density, $\rho = \frac{P}{RT}$ $\qquad P$ – Pressure $\left[\frac{N}{\text{m}^2}; \frac{\text{lb}}{\text{ft}^2}\right]$

(Ideal Gas Law) $\qquad R$ – Gas constant for air $R = 1716\ \frac{\text{ft} \cdot \text{lb}}{\text{Slug °}R}$

$$= \frac{J}{\text{kg} \cdot K}$$

T – Absolute temperature [K, °R]

This assumes that the gases in consideration obey the Ideal Gas Law. Gases under normal pressure and temperature approximate ideal gas behavior.

EXAMPLE PROBLEM 1.2.1.2

A 10-liter tank containing compressed air at 400 kPa is to be shipped on a commercial airline to Nairobi, Kenya. The empty container weighs 45N. Determine the air density

$$\left[\frac{\text{kg}}{\text{m}^3}\right]$$

of the air filling the tank and total weight of the container when full. The temperature is 20°C. The gas constant

$$R_{air} = 287\,\frac{\text{J}}{\text{kg}\cdot\text{k}}$$

and the atmospheric pressure = 101 kPa.

Solution:

Assume that Ideal Gas Law applies.

Therefore, density for air,

$$\rho_{air} = \frac{P}{RT}$$

where P = Absolute pressure
T = Absolute temperature

$$\frac{P}{RT} = \frac{\left[400 \times 1000\ \text{N/m}^3 + 101 \times 1000\ \text{N/m}^3\right]}{\left[287\ J\backslash\text{kg}\cdot\text{K}\right][20 + 273]\ K}$$

Note: $1\ J = 1\ \text{N}\cdot\text{m}$

$$\text{Weight of air} = (\rho_{air})(g) \times \text{volume} = \left(5.96\frac{\text{kg}}{\text{m}^3}\right)\left(9.81\frac{\text{m}}{\text{s}^2}\right)\left(10 \times 10^{-3}\ \text{m}^3\right)$$

$$= 0.58\text{N}$$

$$\text{Total weight of the tank} = 45\text{N} + 0.58\text{N} = 45.58\text{N}$$

1.2.2 Viscosity—Dynamic and Kinematic Viscosity

The viscosity of a fluid is a measure of its internal *stickiness*. All real fluids possess this property. Because a certain amount of energy is needed to overcome the stickiness of a fluid as it flows, viscosity results in loss of energy along the direction of flow. Hence the idea of internal friction. The viscosity of liquids is due to intermolecular forces of attraction between molecules. Intermolecular forces of attraction in gases are weak, and in contrast to liquids, viscous forces in gases are due to molecular activity. Because of this difference, liquid viscosity *decreases* as temperature increases; but the reverse is the case for the viscosity of gases.

$$\mu = \frac{Shear\ stress}{Strain\ rate} = \frac{\tau}{\frac{du}{dy}}$$

Consider the following setup:
A thin film of fluid (i.e., motor oil) of thickness h is between two plates. The lower plate is stationary and the upper plate is sliding at a constant velocity, V. Because of the non-slip condition, the velocity of the fluid in contact with either plate is the same as the velocity of the plate as shown.

Referring to Figure 1.2.2.1, the rate of strain (or strain rate) is equal to the velocity gradient between the plates (assumed to be linear as long as this film is only a few millimeters thick), and is given by

$$\frac{u(h) - u(0)}{h} = \frac{du}{dy} \quad \text{or} \quad \frac{du}{dy} = \frac{u(0) - u(h)}{h}$$

Therefore, shear stress on the walls of either plate $\tau = \mu\left|\frac{du}{dy}\right| = \mu\left|\frac{V}{h}\right|\left[\frac{N}{m^2}\right]$

μ is called the absolute viscosity, but it is also known as coefficient of viscosity, dynamic viscosity, and viscosity. Units of μ are

$$Pa \cdot s \quad \text{or} \quad \left[\frac{N \cdot s}{m^2}\right]$$

in SI units or

$$\frac{\text{lb} \cdot s}{\text{ft}^2}$$

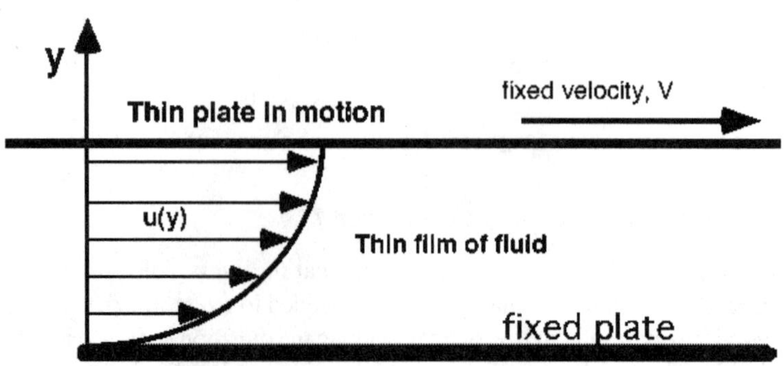

FIGURE 1.2.2.1 Thin Film of fluid between two parallel plates.

in English units. The following table gives other units for absolute viscosity:

Quantities involving dynamic and kinematic viscosity

1 poise = dyn · s/cm^2
1 dyn (a unit of force) = 10^{-5} N

1.	1 N · s/m^2	= 1 Pa · s = 10 poise
2.	1 poise	= 100 centipoise
3.	1 centipoise	= 1/100 poise
4.	1 centipoise	= 1/1000 Pa · s = 1 mPa · s (i.e., 1 mili Pa · s)
5.	1 N · s/m^2	= 1 Pa · s = 1000 centipoise
6.	1000 centipoise	= 1 N · s/m^2 = 1 Pa · s

Classification of Fluids by Viscosity

Fluids that possess a linear relationship between shear stress and strain rate are called *linear* or Newtonian fluids; otherwise, they are classified as non-Newtonian fluids. Besides non-Newtonian fluids that behave in an even more complex manner, there are four common classes of fluids as shown in Table 1.2.2.1, which corresponds to Figure 1.2.2.2.

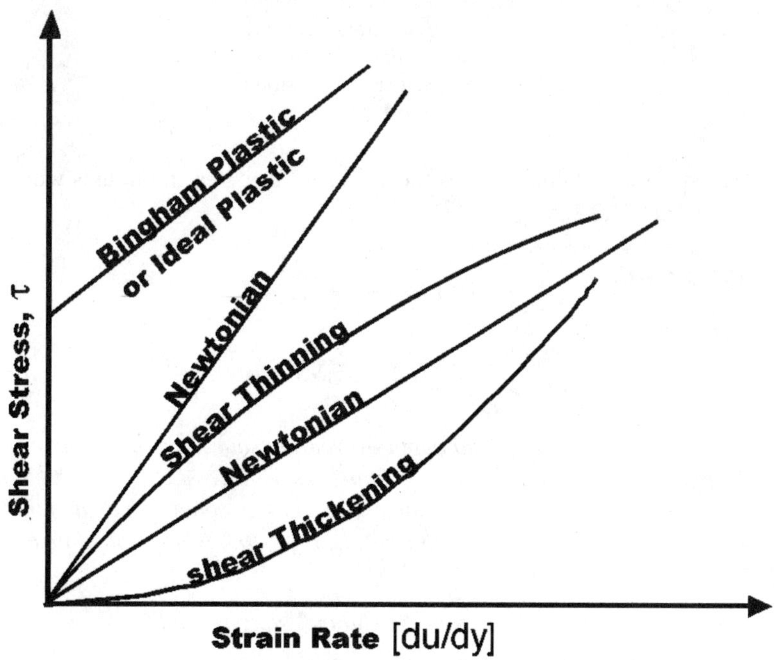

FIGURE 1.2.2.2 Viscosity classification of fluids.

TABLE 1.2.2.1 Classification of Fluids

Classification	Relationship between shear stress and strain rate and behavior	Examples
Newtonian or linear fluids	Linear relationship between shear stress and strain rate	Water, air, and most oils
Shear thickening, also called dilatants	Non-linear relationship between shear stress and strain rate. In this case, as strain rate is increased, shear stress increases more than proportionately.	Quicksand (water-sand mixture), clay slurries, and candy compounds. Pumps that are used to pump these fluids must be select-selected with premium judgment because at high enough shear rates these liquids behave like solids.
Shear thinning, also called pseudoplastic fluids	Non-linear relationship between shear stress and strain rate. Less rapid increase in shear stress as the strain rate is increased.	Latex paint, colloidal suspensions, grease, polymer solutions, most emulsions, and soaps. They present no pumping problems
Bingham plastic, or ideal plastic fluids	These are neither fluids nor solids. In essence, they behave either way depending on the imposed shear stresses. These fluids require a minimum finite shear stress to cause initiate flow (the behavior of solids). Once the yield stress is exceeded, they flow acting like fluids. Once flow is initiated, a linear relationship between shear stress and strain rate is observed.	Toothpaste, solid grease, mayonnaise, some gels, tomato catsup. With regard to pumping, they behave like pseudoplastics.

A few equations relating to dynamic viscosity for gases and liquids with temperature are given in Table 1.2.2.2.

EXAMPLE 1.2.2.1

Let

$$\mu(h) = 2.5(gh_oS_o)^{1/2}\,\text{1n}\left(\frac{34.5h}{\epsilon}\right)$$

represent the velocity u (h) and depth (h) relationship in an open channel. H_o and S_o are the depth of flow on the longitudinal slope, respectively. Let S_o = 0.0065 and h_o = 2.0 m. The nominal or characteristic roughness height ϵ is about 15 mm and the critical shear stress to move the channel bed material, $t_{critical}$ = 0.8 N/m^2. Calculate or estimate the following for water at 25°C:

- *The velocity of flow 50 cm from the channel bottom in m/s*
- *The shear stress at 15 mm from the channel bottom*
- *Determine whether the channel is stable for the given slope, flow depth and ϵ value*

TABLE 1.2.2.2 Equations Relating Dynamic or Kinematic Viscosity with Temperature

Gases	Liquids
1) $\dfrac{\mu}{\mu_o} = \left[\dfrac{T}{T_o}\right]^n$ – power law For air, $n = 0.67$ μ_o = Reference μ @ $T = T_o$ Temperature must be expressed in absolute units (K or °R)	1) $\rho u \dfrac{\mu}{\mu_o} = a + b\left(\dfrac{T_o}{T}\right) + c\left(\dfrac{T_o}{T}\right)^2$ a, b, c are constant values that can be determined if μ at 3 temperature values are known.
2) $\mu = \dfrac{CT^{2/3}}{a+b} = \dfrac{\left(\dfrac{T}{T_o}\right)^{2/3}(T_o + S)}{T+S}$ This is the Sutherland Law. S and C can be found if μ is known at 2 temperatures. Thus S and C are empirical constants. For example, S_{air} = 100K or 199° R	2) Andrade's equation $\mu = K_1 e \dfrac{K_2}{T}$ T = Absolute temperature (K or °R) K_1, K_2 are constants for the liquid found for any 2 temperatures

Solution:

Given:

$$\epsilon = \frac{15}{1000} = 0.015 \text{ m}$$

$$S_o = 0.0065$$

$$h_o = 2.0 \text{ m}$$

$$h = \frac{50}{100} \text{ m} = 0.5\text{m}$$

1. $\mu(h) = 2.5(gh_oS_o)^{1/2}\ln\left(\dfrac{34.5h}{\epsilon}\right)$

$$\mu(h = 0.50) = 2.5[9.81(2.0)(0.0065)]^{1/2}\ln\left[\frac{34.5(0.50)}{0.015}\right]$$

$$= 6.29 \text{ m/s}$$

At any h, $\mu(h) = 0.8928\ln\left(\dfrac{34.5h}{0.015}\right) = 0.8928\ln(2300h)$

2. $\tau = \mu\left|\dfrac{du}{dy}\right| = \mu\dfrac{du}{dh} = \mu\left(0.8928\dfrac{2300}{2300h}\right) = 0.8928\dfrac{\mu}{h}$

NOTE: $\dfrac{d(\ln u)}{dx} = \dfrac{1}{u}\dfrac{du}{dx}$

μ at 25° $C = 8.9 \times 10^{-4}\dfrac{N \cdot s}{\text{m}^2}$; h = 15 mm = 0.015 m

$$\therefore \tau = \frac{0.8928 \times 8.90 \times 10^{-4}}{0.015} = 0.053\frac{N}{\text{m}^2}$$

3. The critical shear stress = 0.8 N/m^2 but the shear stress at h = 15mm = 0.053 N/m^2.

Therefore, the channel will not erode. It is stable.

Kinematic viscosity (ν) is defined as the dynamic or absolute viscosity divided by the fluid density. Thus:

$$\nu = \frac{\mu}{\rho}\left[\frac{m^2}{s}, \frac{ft^2}{s}\right]$$

At 20°C

$$\rho \text{ of water} = 998 \text{ kg/m}^3$$

$$\mu \text{ of water} \cong 1.00 \times 10^{-3} \frac{N \cdot s}{m^2}$$

$$= 1.00 \times 10^{-3} \text{ Pa} \cdot \text{s}$$

$$= 1.00 \text{ mPa} \cdot \text{s}$$

$$= 1 \text{ centipoise}$$

$$\therefore \nu = \frac{1 \times 10^{-3} \frac{N \cdot s}{m^2}}{998 \frac{kg}{m^3}} \cong 1 \times 10^{-6} \text{ m}^2/\text{s}$$

1.2.3 Surface Tension

Surface tension, a property possessed only by liquids, results from the unbalanced attractive forces between molecules at the surface of a liquid. This occurs at liquid-gas interface or a liquid-liquid interface for liquids that do not mix (immiscible liquids). Surface tension forces cause a liquid to develop a skin-like surface, which can be evidenced by the following, often observed phenomena:

- Formation of bubbles
- Formation of droplets
- Needle made of material denser than water floats
- Small drops of mercury form beads on glass
- Capillary rise of water in small glass tubes and capillary fringe above water table in soils
- Capillary depression of mercury in a capillary tube.

Surface tension (σ) of a liquid (also called coefficient of surface tension) is the *force* per unit length acting in the surface at right angles to one side of a line drawn in the surface. Units are

$$\frac{N}{m} \quad \text{or} \quad \frac{lb}{ft}$$

Force due to surface tension = $\sigma \times$ length of contact.

Surface Tension Phenomena. Forces of adhesion are attractive forces between molecules that are dissimilar. Cohesive forces are forces of attraction between like molecules.

With respect to adhesive and cohesive forces, the concept of angle of contact can be defined:

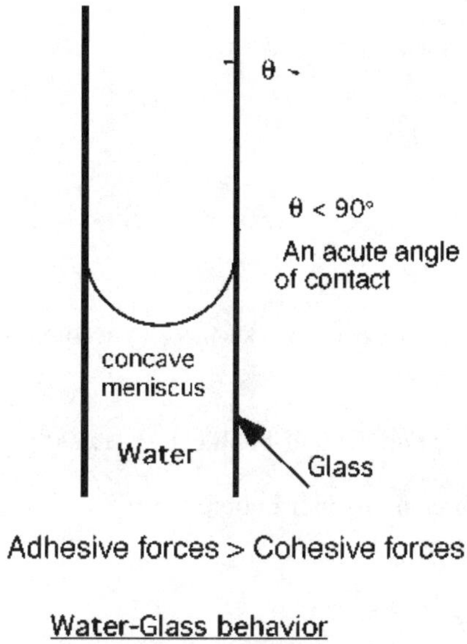

FIGURE 1.2.3.1a Water-glass behavior.

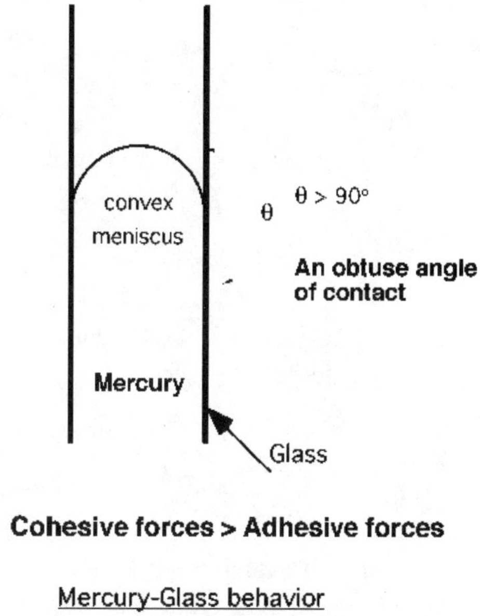

FIGURE 1.2.3.1b Mercury-glass behavior.

Droplets

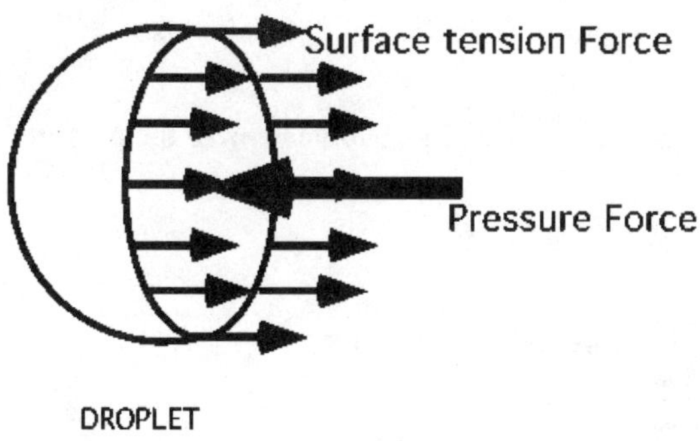

FIGURE 1.2.3.2 Forces in a bubble.

In the above figure, $\pi D\sigma$ is equal to the surface tension force in the half droplet. And

$$\frac{p\pi D^2}{4}$$

is the pressure force in the droplet as shown

where D = Diameter of droplet
σ = Surface tension of the water.

Therefore

$$\pi D\sigma = \frac{p\pi D^2}{4}$$

And the pressure in the droplet,

$$P = \frac{4\sigma}{D}$$

Bubbles Since bubbles are like shells, there are two surfaces to account for in contrast to droplets. Thus:

$$\text{Surface tension force on the half-bubble} = 2 \times \pi D\sigma$$

$$\text{and pressure force in the half-bubble} = \frac{P\pi D^2}{4}$$

Therefore

$$2 \times \pi D\sigma = \frac{P\pi D^2}{4}$$

Simplifying, pressure in bubble,

$$P = \frac{8\sigma}{D}$$

For the same surface tension, the pressure in a bubble is *twice* the pressure in a droplet for the same conditions.

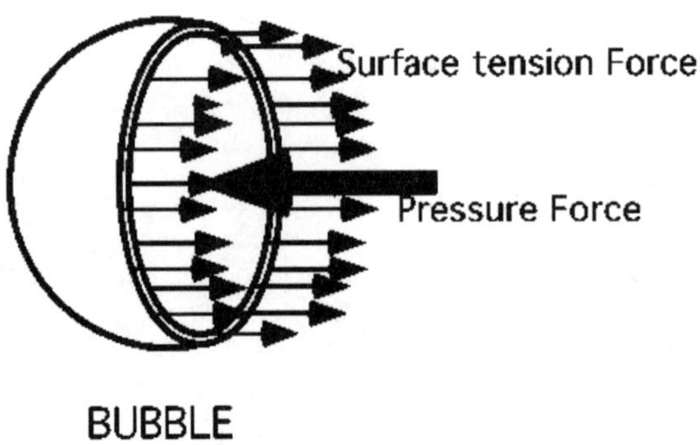

FIGURE 1.2.3.3 Forces in a bubble.

Surface-tension forces can be significant in a number of situations in the mechanics of fluids involving liquid-gas, liquid-gas-solid, or liquid-liquid interfaces:

- Transport of liquid through porous media (surface tension forces can be significant in groundwater problems)
- Flow of thin films and very shallow depth flows (sheet flows). In the development of scale models, the modeler must be careful to account for surface tension forces, which otherwise might dominate in the scaled-down model as opposed to the prototype.
- Formation of bubbles and droplets
- Breakup of jets
- Capillary flows, as mentioned previously.

EXAMPLE 1.2.3.1

Calculate the capillary rise of water in a 1 mm glass tube for water at 20°C and compare it with the depression that would occur if mercury replaces the water at the same temperature.

TABLE 1.2.3.1 Surface Tension (in contact with air) at 20°C and 25°C for Common Liquids

Liquid	Surface tension [N/m]	
	20°C*	25°C†
Ethyl alcohol	0.0228	0.0223
Methyl alcohol	0.0230	0.0222
Benzene	0.0290	0.0282
Bromine	0.0440	—
Carbon disulphide	0.0336	0.0323
Carbon tetrachloride	0.0269	0.0263
Chloroform	0.0272	0.0271
Ether	0.0165	0.0164
Glycerine	0.0633	0.630
Mercury	0.466	—
Olive oil	0.0320	—
Paraffin oil	0.0264	—
Toluene	0.0288	—
Turpentine	0.0273	—
Water	0.0728	0.07197

* Most values from Yarwood & Castle (1970). Physical and mathematical tables, Macmillan.

† Most values from CRC Handbook of Tables for applied Science (1973) 2nd Ed., CRC Press.

TABLE 1.2.3.2 Angles of Contact for Selected Liquids Between Glass and Liquids

Liquid	Angle of Contact [°]*
Water (very clean glass)	0
Alcohol (very clean glass)	0
Chloroform	0
Ether	16
Turpentine	17
Paraffin oil	26
Mercury	127
Mercury (fresh drop)	139

* From Yarwood & Castle (1970), Physical and Mathematical Tables, Macmillan.

Solution:

Capillary rise or depression

$$h = \frac{4\sigma \cos\theta}{\gamma D}$$

For water:

$\theta = 0°$

$\sigma = 0.0728$ N/m

$\gamma = 9789$ N/m^3

$$h = \frac{4(0.0728)\cos 0°}{(9789)(0.001)} = 0.030 \text{ m} = 30 \text{ mm (rise)}$$

For mercury:

$\theta = 127°$

$\sigma = 0.465$ N/m

$\gamma = 132800$ N/m^3

$$h = \frac{4(0.466)\cos 127°}{(132800)(0.001)} = -0.0084 \text{ m} = 8.4 \text{ mm (depression)}$$

1.2.4 Vapor Pressure

Liquids, in general, will evaporate when exposed to the atmosphere. After a rainfall event, water dries from the pavement, roof, and other forms of cover. It also is true that dew collects on vegetation (especially on grassed lawns) overnight. These phenomena are linked to the concept of vapor pressure. Water boils when heated to a certain temperature (usually about 100° C), but it tends to boil more quickly (usually at a temperature lower than 100° C) at higher altitudes. We also notice that evaporation increases when the atmosphere is drier (when the relative humidity is low).

When a liquid is exposed to surrounding air or vacuum, its molecules translate from liquid state to gaseous state to fill the space immediately above the liquid surface. If this space is limited or the movement of air is restricted, a situation is reached where there is a net zero translation of molecules from gaseous state to liquid state, and vice versa. At this state of equilibrium, the partial pressure exerted by the liquid is called its vapor pressure, or more accurately *saturation vapor pressure*.

Saturation vapor pressure (SVP) depends on the level of energy of liquid molecules, and therefore is directly dependent on the temperature of the liquid. For example, the SVP of water at 20° C is 2.34 kPa, and at 100° C it is 101.3 kPa. The SVP of water at 100° C is the same as the standard barometric pressure. Thus, a liquid such as water boils when its SVP is exactly equal to the ambient pressure. If the ambient pressure is lowered, water, for example, can be made to boil at temperatures lower than 100°C. Therefore, it is observed that water also boils at temperatures lower than 100°C in higher altitudes where the ambient pressure is lower than 101.3kPa.

Cavitation

Cavitation is the condition in which the pressure in a liquid is lowered to match its vapor pressure, forming vapor cavities within the liquid body. For example, this condition can be induced when flow of water goes through a constriction or obstruction in the flow path. As the water passes through a smaller area of flow, its velocity must increase to maintain the same flow rate. The increase in velocity, however, results in a lowering of the pressure in the liquid. Cavitation occurs if the pressure in the liquid is lowered *below* the SVP at the temperature of the liquid. The vapor cavities contain water vapor at a pressure that matches the SVP of the liquid at the ambient (liquid) temperature. Thus, the liquid is said to boil. In the case of water, visible bubbles form in the body of the liquid.

Cavitation is a concern in constrictions in conduit flow (around valves) and in pumps. For example, in water, vapor bubbles that develop in the flow (during cavitation) are swept downstream to regions of higher liquid pressure. There they collapse rapidly and can do so with enough intensity to cause structural damage due to *pounding* that occurs because of the large pressure transients that develop. Thus, cavitation can cause severe damage in piping systems and to rotors and blades of pumps and turbines.

EXAMPLE 1.2.4.1

Determine the temperature at which water will boil in a mountainous region at an altitude of 4000 m. Use information in the Appendix tables—Standard Atmospheric Properties and Properties of Water.

Solution:

At an altitude of 4000 m, atmospheric pressure = 61.64 kPa.
For water to boil, its SVP must be 61.64 kPa. Examining the table on Properties of Water:

T°C	SVP [kPa]	
80	47.3	
90	70.1	
?	61.64	pressure must match atmospheric pressure

By linear interpolation:

$$\text{Boiling temperature,}\ T_{Boil} = 80° + \frac{[90 - 80]}{[70.1 - 47.3]}(61.64 - 47.3)$$

$$T_{Boil} = 86.3°\text{C}$$

This demonstrates that water will boil at a lower temperature if the ambient pressure is lowered. In fact, water can boil at room temperature if the ambient pressure is lowered to the SVP of water at the room temperature.

1.2.5 Compressibility of Fluids

The bulk modulus (E_v) of a fluid is used to characterize its compressibility. The bulk modulus also is referred to as the volume modulus of elasticity and the coefficient of compressibility. It has the same units as pressure

$$\left[\frac{N}{\text{m}^2} \text{ or } \frac{\text{lb}}{\text{ft}^2}\right]$$

—the pressure change that is required to change the volume of the fluid by 100 percent at constant temperature:

$$\text{Thus,} \qquad E_v = \frac{d_p}{\frac{d_\forall}{\forall}} = \frac{d_p}{\frac{d_\rho}{\rho}}\left[\frac{N}{\text{m}^2}; \frac{\text{lb}}{\text{ft}^2}\right]$$

It also is called the coefficient of compressibility.

To gain an appreciation for the compressibility of water, consider that its bulk modulus is 2.1×10^6 kPa, which is approximately 2100 atmospheres (1 atmosphere = 100 kPa). Thus, to change 100 m^3 of water by 1 m^3 or 1 percent, it would require 21,000 kPa. This shows that water is virtually incompressible. Therefore, liquids such as water are considered incompressible for most day-to-day engineering problems.

It must be noted, however, that the compressibility of liquids is not ignored completely. For example, the water hammer phenomenon in piping systems is due to the compressibility of water as very large pressure changes occur in a pipe due to such events as abrupt valve closure and opening.

In general, if changes in density are 4 percent or more, the fluid is considered compressible, while changes less than 3 percent are considered negligible.

The speed of sound in a liquid is calculated by the following relationship, which involves the bulk modulus:

$$\text{Speed of sound} = \sqrt{\frac{E_v}{\rho}} \quad \text{where } \rho = \text{density of the liquid.}$$

The speed of sound (in a fluid), also referred to as the *acoustic velocity*, is the velocity at which small disturbances propagate in a fluid. The compressibility of gases is a little more complex because other processes are involved.

An *isothermal* process is one in which compression or expansion occurs at constant temperature. From ideal gas law, $\frac{P}{\rho}$ constant and $E_v = P$. An *isentropic* process is one in which compression or expansion occurs without friction and there is no heat exchange with the surroundings:

Here $\frac{P}{\rho^k}$ = constant, where $k = \frac{C_p}{C_v}$ and $E_v = P$; where C_p and C_v are the specific heats of the gas at constant pressure and volume, respectively.

The specific heats for gases are discussed in section 1.2.6 below.

1.2.6 Specific Heat and Latent Heat

The specific heat of a substance is the heat (energy) required to raise a unit mass of the substance by 1 degree. Units are

$$\frac{\text{cal}}{\text{g} \cdot {}^\circ\text{C}} \quad \text{or} \quad \frac{\text{J}}{\text{kg} \cdot \text{k}} \quad \text{or} \quad \frac{\text{Btu}}{\text{slug} \cdot {}^\circ\text{R}}$$

For gases in particular, the process by which heat is added to a substance is important, and therefore, there are two specific heat values: specific heat at constant pressure (C_p) and specific heat at constant volume (C_v). The ratio of these specific heats is

$$k = \frac{C_p}{C_v} \geq 1.00$$

(usually greater than 1), which implies that $C_p \geq C_v$. Furthermore, the gas constant, R, has the following relationship with C_v and C_p:

$$R = C_p - C_v.$$

1.2.7 Ideal Gas Law

$$P = \rho RT$$

where P = Pressure
ρ = Density
T = Absolute temperature
R = Ideal gas constant for a particular gas or mixture of gases (e.g., air).

In addition, R is related to the universal gas constant, R_u as follows:

$$R = \frac{R_u}{M_{gas}}; \qquad R_u = 8314 \frac{J}{kg - \text{mol} \cdot k}$$

$$= 49710 \frac{\text{ft} - \text{lb}}{\text{slug} - \text{mol} \cdot {}^\circ\text{R}}$$

where

$$M_{gas} = \text{Molar mass of the gas} = \left[\frac{\text{kg}}{\text{kg} - \text{mol}}\right]$$

With respect to the specific heats (C_v and C_p) previously discussed in Section 1.2.6, we can develop the following relationships for the internal energy (or change thereof) and enthalpy (or changes thereof).

Internal Energy (U)

The change in the internal energy $U_2 - U_1$ [in a flow from section (1) to section (2)] is related to a change in temperature because the internal energy of an ideal gas depends on its temperature.

Therefore, $U_2 - U_1 = C_v(T_2 - T_1)$, where C_v is assumed constant for moderate change in temperature, $T_2 - T_1$.

Change in Enthalpy (*h*)

The change in enthalpy in a flow from section (1) to section (2) is given by:

$$h_2 - h_1 = C_p(T_2 - T_1)$$

where, as before, C_p is assumed to be constant for a fairly moderate change in temperature.

CHAPTER 2
FLUID STATICS

This chapter deals with fluid statics—the forces exerted in a motionless fluid. In particular, it covers the concept of pressure, pressure distribution, forces on submerged surfaces—including plane and curved surfaces. Forces on dams are examined by providing solved examples. Basic principles in manometry are developed, applied and discussed. Archimedes' principle is stated and discussed to show how buoyant forces on partially or completely submerged objects may be calculated. Pascal's law is presented and its application to a simple system is explored by solving an example problem.

2.0 FLUID STATICS

In this section, several topics that can be termed *hydrostatics* are presented—including the concept of pressure due to fluids. Several facets of hydrostatics, and the engineering applications of these principles, are summarily described. The intent is not to provide a detailed discussion of these topics, but to provide a short review of the key concepts.

2.1 CONCEPT OF PRESSURE

$$\text{Pressure} = \frac{\text{Normal pressure}}{\text{Area normal to force}} \left[\frac{\text{N}}{\text{m}^2} \text{ or } \frac{\text{lb}}{\text{ft}^2}\right]$$

In English units, pressure is normally given in pounds per square inch (psi) or pounds per square foot (psf) but in SI units the Pascal (Pa) is used:

$$1 \text{ Pa} = 1 \frac{\text{N}}{\text{m}^2}$$

Pressure is a scalar quantity, which means it does not have an associated direction. The pressure at a point is the same in all directions and depends only on the vertical distance from a reference point in the fluid(s). Pressure increases with a decrease in elevation, and vice versa. Pressure forces on a surface are normal to the surface.

Important Concepts

1. Pressure at a point is the same in all directions.

$$P_A = P_B = P_C$$

2. Pressure forces act perpendicularly on a surface.

Theoretically, change in pressure with elevation (or vertical height z) is given by:

$$\frac{dp}{dz} = -\gamma$$

where γ = Specific weight of the fluid depending on the situation; γ could be a function of several factors (but usually z and pressure).

2.2 PRESSURE SCALES

Figure 2.2.1 presents the usual scales used when pressure is measured. Indeed, pressure can be expressed with reference to any arbitrary datum.

Pressure is one of the fundamental quantities of interest in hydraulics, and because of that importance, many ways have been developed to measure it. Pressure is measured with reference to either of two pressure scales or reference points. If the reference is to absolute zero pressure (complete vacuum), then the reported pressure is said to be *absolute* pressure. Barometric pressure is measured as absolute pressure.

If pressure is referenced to the prevailing (surrounding) or atmospheric pressure, it is called *gage* pressure. Often the pressure reported in conduits flowing under pressure is gage pressure. When one measures the pressure in an automobile tire, it is gage pressure. Gage pressures can be negative, but absolute pressures are always positive.

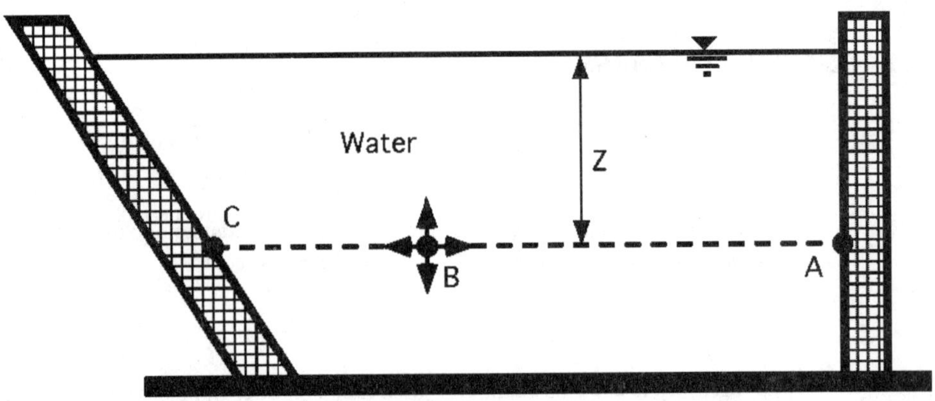

FIGURE 2.1.1 Pressure is the same in all directions.

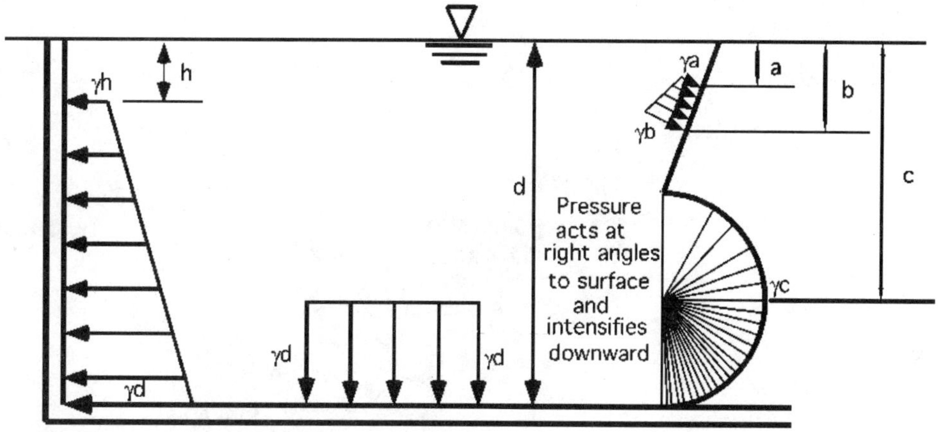

FIGURE 2.1.2 Pressure distribution on submerged surfaces.

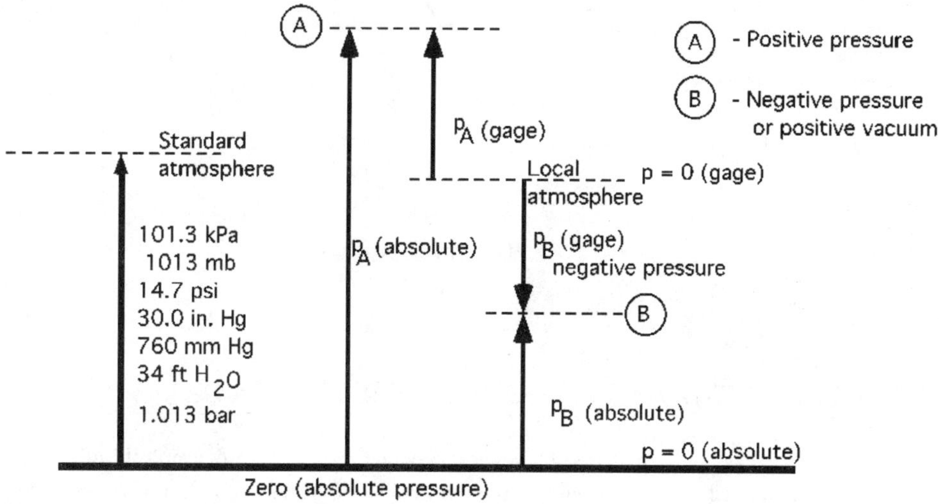

FIGURE 2.2.1 Gage and absolute pressure scales.

The standard pressure of the atmosphere is 14.696 psi (101.325 kPa) at sea level. The pressure in the atmosphere at a given location is due to the weight of the column of air above that particular point.

$P_{absolute} = P_{gage}$ + Atmospheric pressure with reference to Figure 2.1.1

So also: $P_1 = P_2 + P_{atmospheric}$

For example, if $P_1 = 187$ kPa (absolute),

then $P_2 = (187 - 101.325)$ kPa (gage)

$= 85.675$ kPa (gage)

Again, absolute pressure is always positive (never less than 0.0).

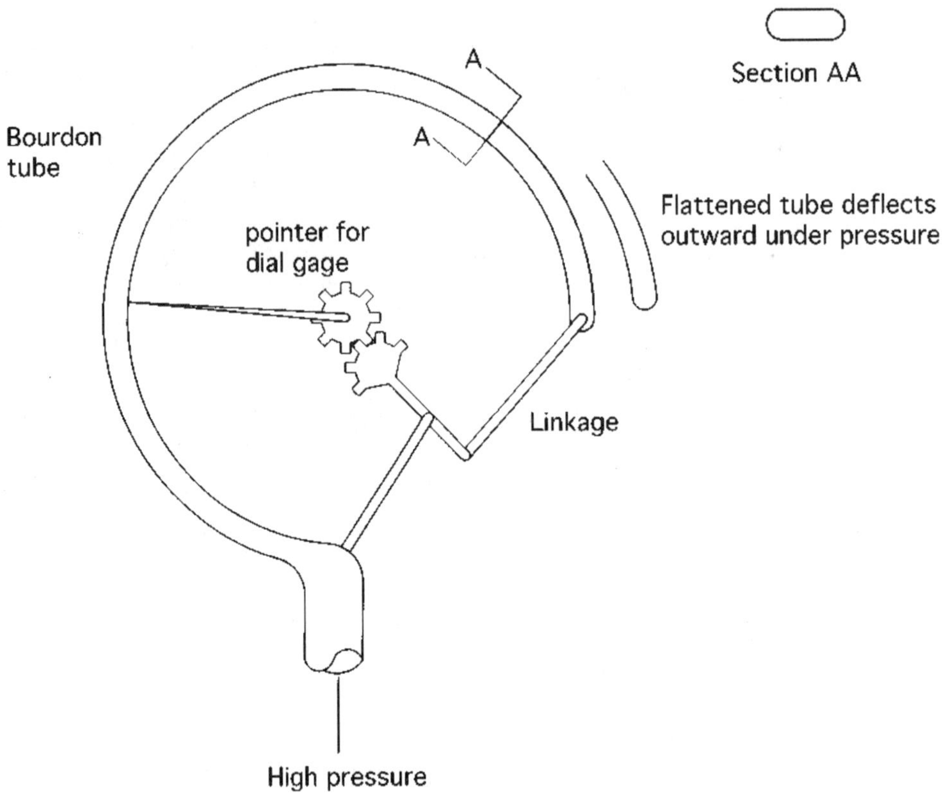

FIGURE 2.3.1 Bourdon gage.

2.3 MANOMETRY AND PRESSURE MEASUREMENT

Pressure is measured by several devices, ranging from simple U-tube manometers to very sophisticated electronic pressure transducers, which can measure very small pressure changes. In general, atmospheric pressure is measured by barometers and therefore is referred to as barometric pressure.

Gage pressure can be measured by U-tube or differential manometers and a variety of other types of pressure gages. The following table provides brief descriptions of several types of pressure and barometers. It should be noted that the use of mercury barometers is becoming more restricted because of the health hazards posed by vapors from broken instruments.

Use of U-tube (differential) Manometers

Consider the simple U-tube manometer shown in Figure 2.3.2.

It should be noted that by first principles, the pressure at [2] and the pressure at [3] are equal. This is because the pressure at a point depends only on its vertical distance from a reference point and the manometer fluid (actually the specific weight of the fluid). Here,

TABLE 2.3.1 Types of pressure gages and barometers

Barometers The two common types of barometers are mercury and aneroid.	Barometers are used to measure atmospheric pressure. When barometric pressure is measured, the value obtained is on the absolute pressure scale. Mercury barometers utilize a column of mercury as the manometer fluid. The aneroid barometers utilize a system of mechanical linkages connected to an evacuated chamber that deforms in response to changes in pressure. The response to change in pressure is transmitted through linkages connected to the dial.
Manometers • Can be used to measure large pressure differences • Small pressures	See the previous and other descriptions in this section.
Bourdon tube gage • Easy to use and very convenient. • Not used for very accurate pressure measurements or to measure small pressure changes. • Often requires calibration.	Is a common mechanical gage made of a hollow metal tube which is designed to lengthen or shorten due to pressure changes (See Figure 2.3.1). The *flexing* tube is connected by linkages to an appropriately graduated dial.
Pressure transducers: Several types, including: • Electronic • Condenser microphone • Piezoelectric pressure transducer • Strain gage transducer (for very accurate measurement of large, small, dynamic, and static pressure)	Can be used to measure static, dynamic, small, and large pressure changes, and absolute pressures.

the fluid in the column from [1] to [2] has a specific weight of γ_1 and the fluid from [2] to [3] has a specific weight equal to γ_2. Since points [2] and [3] are in the same fluid and at equal elevation, $P_2 = P_3$. If it is assumed that the pressure at [1] is atmospheric, then the following relationships can be inferred:

$$P_2 = P_1 + \gamma_1 z$$
$$P_2 = P_3 \quad \text{(by definition)}$$
$$\therefore P_3 = P_1 + \gamma_1 z$$

Applying the same (simple) principles to a more complex manometer arrangement, refer to Example 2.3.1, the above manometer principles can be used to effectively analyze a more realistic manometer arrangement.

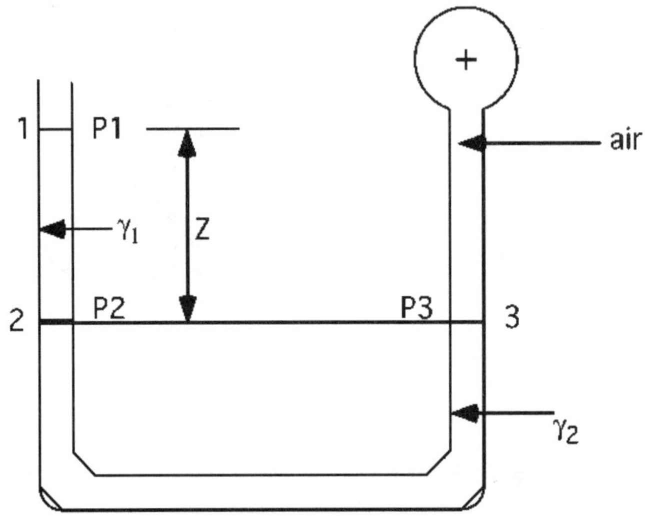

FIGURE 2.3.2 Simple U-tube manometer.

EXAMPLE 2.3.1

For the manometer arrangement given below, determine the pressure difference in kPa between points A and B.

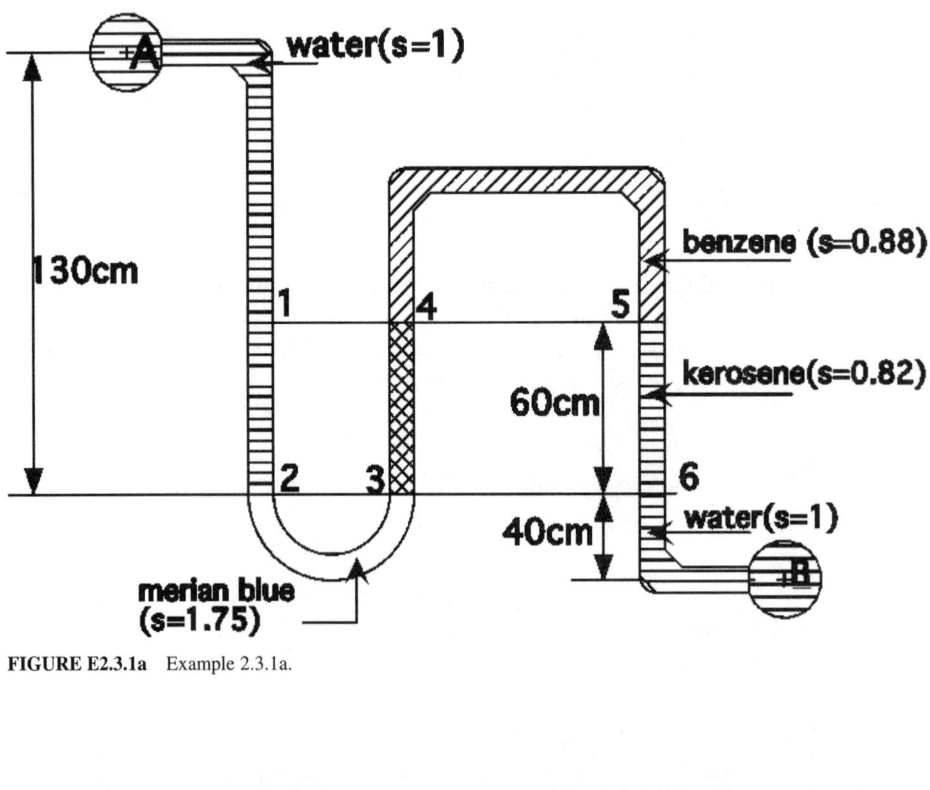

FIGURE E2.3.1a Example 2.3.1a.

Solution:
First, observe that $P_2 = P_3$ and $P_4 = P_5$, but $P_1 \neq P_4$ and $P_3 \neq P_6$.
In the figure, S is the specific gravity of the fluid in the manometer as shown.

$$P_B = P_A + \frac{130}{100}\gamma_{water} - \frac{60}{100}\gamma_{MB} + \frac{60}{100}\gamma_{\text{ker }o} + \frac{40}{100}\gamma_{water}$$

$$\gamma_{water}\ @4°C = 9810\,\frac{N}{m^3}\ \therefore\ \gamma_{water} = (S_{water}) = 1 \times 9810\,\frac{N}{m^3}$$

$$\begin{aligned} P_B - P_A &= 1.3(9810) - 0.6(1.75)(9810) + 0.6(0.88)(9810) + 0.4(9810) \\ &= 12753.0 - 10300.5 + 5179.7 + 3924 \\ &= 11556.2 \text{ Pa} = 11.56 \text{ kPa} \end{aligned}$$

EXAMPLE 2.3.2

What is the pressure at the center of pipe B in the inclined manometer setup shown below? Give your answer in kPa gage. Assume that $\gamma^1 = 12$ kN/m^3 and $\gamma^2 = 40$ kN/m^3. (Note: Inclined manometer arrangements are used to improve the accuracy of pressure measurements because the inclined orientation increases the least count of the measuring instrument.)

Solution:

$$P_A = 0$$

$$P_B = 0 + \left[\frac{60}{100} \times \frac{3}{5} - \frac{15}{100}\right] \times y_2 - \frac{50}{100}y_1$$

$$P_B = 0 + \left[\frac{60}{100} \times \frac{3}{5} - \frac{15}{100}\right](40 \times 10^3) - \frac{50}{100}(12 \times 10^3)$$

$$\therefore\ P_B = 0 + 8400 - 6000 = 2400 \text{ Pa} = 2.4 \text{ kPa}_{gage}$$

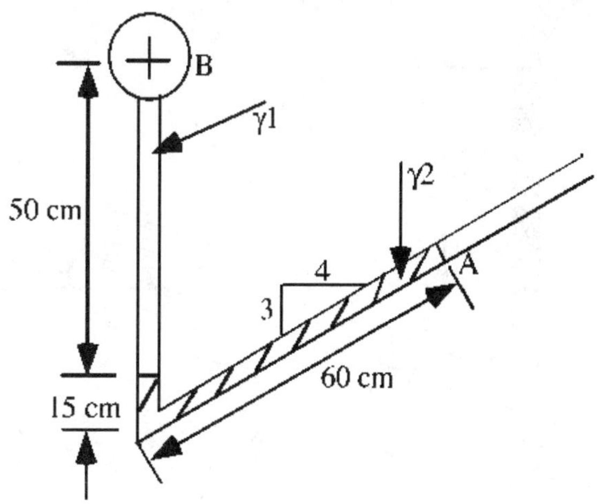

FIGURE E2.3.2a Manometer arrangement for Example 2.3.2.

EXAMPLE 2.3.3

For the manometer arrangement given below in Figure E2.3.3 (a), x = 140 cm and y = 50 cm. The elevations Z_E and Z_F are 1.7 m and 3.2 m, respectively. Determine:

1. The pressure difference [kPa] between points E and F, $P_F - P_E$ in kPa
2. P_F in the following units when $P_E = 45$ kPa and the atmospheric pressure for the mountainous area is 95 kPa:
   a. kPa_{gage}
   b. $kPa_{absolute}$
   c. Meters of water (gage pressure)
   d. Centimeters of mercury (specific gravity = 13.6)

Solution:

1. For water at 40°C, the specific weight, $\gamma_{water} = 9730\ N/m^3$
   Let γ be the specific weight of water at 4°C = 9810 N/m^3
   $Fw = (\gamma\bar{h})(Area)$
   $Fw = (\gamma\bar{h})(Area)$
   $Fw = (\gamma\bar{h})(Area)$
2. **a.** For $P_E = 45$ kPa
   Since $P_F - P_E = -13.57$ kPa
   $P_F = P_E - 13.57 = 45 - 13.57 = 31.43\ kPa_{gage}$
   b. Since [from 2-a] $P_F = 31.443\ kPa_{gage}$
   $P_{F\,(absolute)} = P_{F\,(gage)} +$ atmospheric pressure
   $P_F = 31.43 + 95 = 126.43\ kPa_{absolute}$

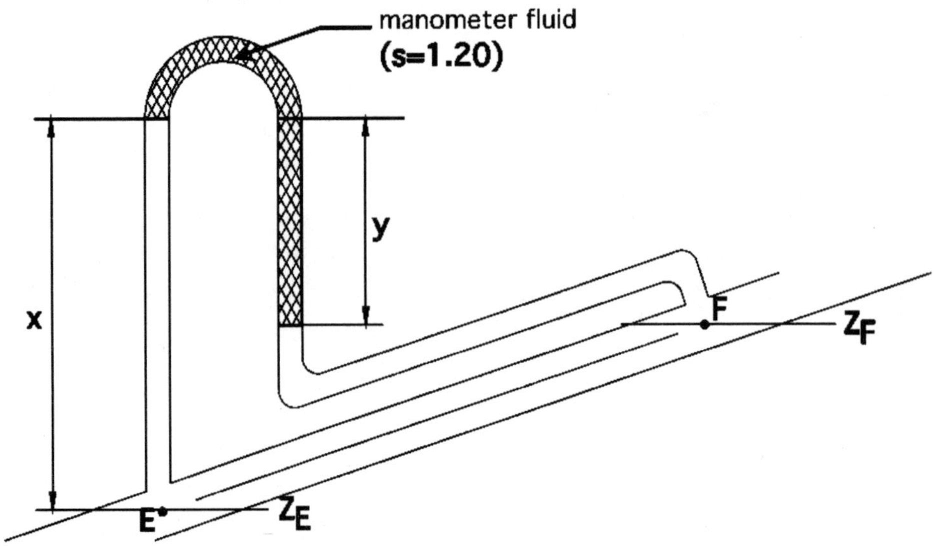

FIGURE E2.3.3a Diagram for Example 2.3.3.

c. P_F in meter of water is determined as follows:

$$\gamma_{water} \text{ @4°C} = 9810 \frac{N}{m^3}$$

But, Pressure = Specific weight × Vertical distance

$$\text{Vertical distance} = \frac{\text{Pressure}}{\text{Specific weight}}$$

$$\text{Therefore: Pressure (m of water)} = \frac{31.43 \times 10^3 \frac{N}{m^2}}{9810 \frac{N}{m^3}} = 3.204\text{m (gage)}$$

d. To determine the equivalent pressure in cm of mercury, we use the same logic as in the previous problem [2c]

$$\text{cm of mercury} = 100 \frac{cm}{m} = \frac{31.43 \times 10^3 \frac{N}{m^2}}{13.6 \times 9810 \frac{N}{m^3}} = 23.56 \text{ cm of mercury (gage)}$$

EXAMPLE 2.3.4

The air pressure in the closed rectangular tank shown in Figure E2.3.4 is 153.32 $kPa_{absolute}$. The atmospheric (ambient) pressure is 101.32 kPa. The bottom dimension of the tank is 1.2 m by 1.8 m, and a = 0.25 m, b = 0.50 m and c = 0.75 m. The specific gravity of the oil is 0.85.

1. Calculate the total force on the bottom of the tank in kN.
2. Suppose the tank is placed in the cargo bay of a space shuttle, such as Discovery, and the tank is exposed during a space walk when the ambient pressure is 2 kPa.

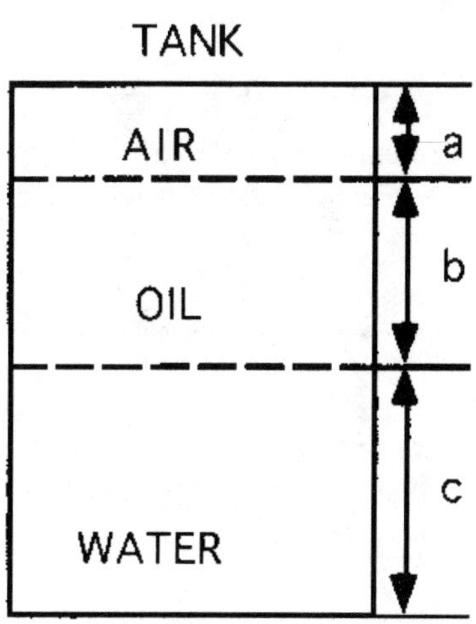

FIGURE E2.3.4a Diagram for Example 2.3.4.

Determine the force on the bottom of the tank (in kN) if the specific weight of the fluids remains unchanged.

Solution:
To solve this problem, the gage pressure at the bottom of the tank P_{bottom}, needs to determined first. Second, the force, F_{bottom} at the bottom can be found by multiplying the area of the bottom of the tank by the pressure at the bottom of the tank.

1. Note that Gage pressure = Absolute pressure – Atmospheric pressure.

Therefore, $P_{air} = 153.32 \text{ kPa} - 101.32 \text{ kPa} = 52.0 \text{ kPa}$

$$P_{bottom} = P_{air} + 0.5 \times 0.85 \times \gamma_{water} + 0.75 \times \gamma_{water}$$

$$\gamma_{water} \text{ @}4°\text{C} = 9810 \frac{\text{N}}{\text{m}^3}$$

$$P_{bottom} = 52 \times 10^3 \frac{\text{N}}{\text{m}^3} + 0.5 \times 0.85 \times 9810 \frac{\text{N}}{\text{m}^3} + 0.75 \times 9810 \frac{\text{N}}{\text{m}^3}$$

$$\therefore P_{bottom} = 63.53 \times 10^3 \frac{\text{N}}{\text{m}^3} = 63.53 \text{ kPa}_{\text{gage}}$$

$$P_{bottom} = \text{Pressure} \times \text{Area} = 63.53 \times 103 \frac{\text{N}}{\text{m}^2} (1.2 \text{ m} \times 1.8 \text{ m})$$

$$= 137.22 \text{ kN}$$

2. If the ambient pressure changes to 5 kPa (in space),

$$P_{bottom} = (153.32 - 5) \times 10^3 \frac{\text{N}}{\text{m}^2} + 0.5 \times 0.85 \times 9810 \frac{\text{N}}{\text{m}^3} + 0.75 \times 9810 \frac{N}{\text{m}^3}$$

$$P_{bottom} = 159.85 \text{ kPa}$$

3. And $F_{bottom} = 159.85\,(1.2 \times 1.8) = 345.27 \text{ kN}$ (much higher than before)

2.4 PASCAL'S LAW

Blaise Pascal (1623-1662) provided contributions in the areas of the principles of the barometer, the hydraulic press and in the subject of interest here—the concept of transmissibility of pressure in a fluid. Pascal's Law states that:

> A change in the pressure applied to an enclosed fluid is transmitted *undiminished* to every portion of the fluid and to the walls of the confining vessel.

This means that pressure changes in a confined fluid are transmitted equally to every portion of the fluid and to the walls of the confining container. This fact has very important implications, as the example below illustrates.

EXAMPLE 2.4.1

Consider the following arrangement of pistons:

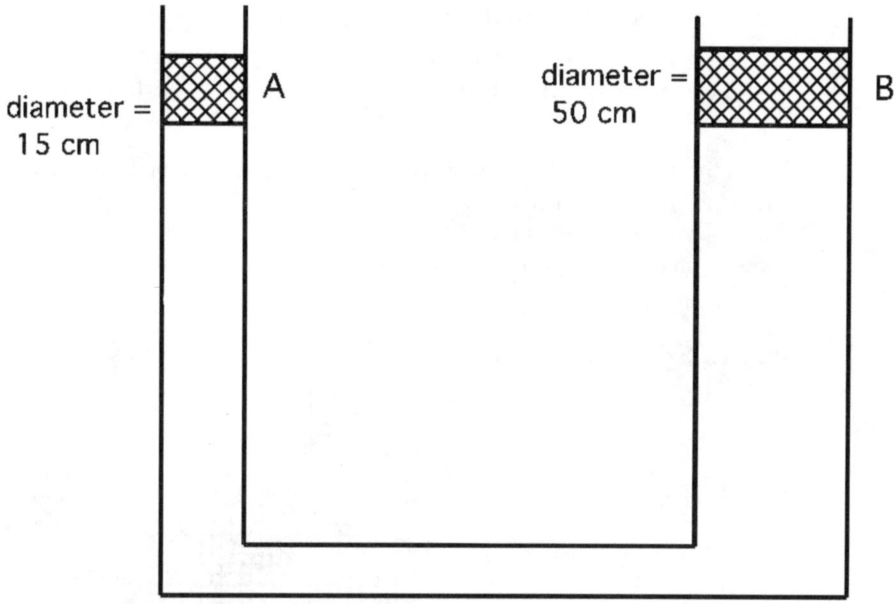

FIGURE E2.4.1a Piston arrangement for Example 2.4.1.

A pressure increase of 210 kPa was applied to the small (A) piston when the pressure at the larger (B) piston was found to be 50kPa. Calculate the force at the larger piston B after the increase in pressure at A was observed.

$$\text{Pressure at B} = (50 + 210)\text{ kPa} = 260\text{ kPa}$$

$$\text{Force at B} = (\text{Area of B})\,(\text{Pressure at B})$$

$$= \left[\frac{\pi(0.5)^2}{4}\right][260\text{ kPa}]$$

$$= [0.19635][260\text{kPa}] = 51.05\text{kN}$$

2.5 FORCES ON A SUBMERGED SURFACE

When a surface is submerged in a fluid, pressure forces act on the submerged surface because of the weight of the fluid above (or on top of) the surface. Thus atmospheric pressure, which can be measured, is due to the column of air above a particular point of interest. For example, if a flexible metal chamber is evacuated by pumping air out of it, the walls begin to collapse (depending, of course, on how flexible it is) because pressure forces (due to atmospheric pressure) now become evident because the counterbalancing pressure formerly inside the chamber has been removed due to the evacuation.

Notable facts about pressure forces:

1. Pressure acts perpendicularly on a surface, and therefore the resultant force due to pressure acts perpendicularly on plane surfaces. Furthermore, if a surface is circular or part its shape is circular, the pressure force acts along the radius of the circle.

2. If the fluid is incompressible, the pressure gradient in the fluid is linear.
3. If pressure gradient is linear and triangular, the location of the center of pressure for a plane surface per unit width, is $\frac{1}{3}$ from the base of the triangular pressure distribution. These concepts are illustrated in Figure 2.5.1.

In the same way, the location of the center of pressure on a submerged surface can be determined by first principles.

When a surface is submerged in a fluid in such a way that the fluid surrounds it, no net pressure forces are experienced by the object.

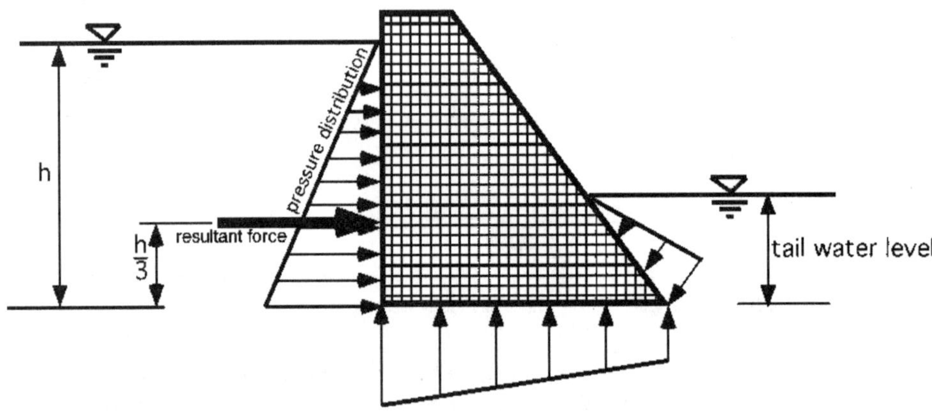

FIGURE 2.5.1 Pressure distribution on a dam.

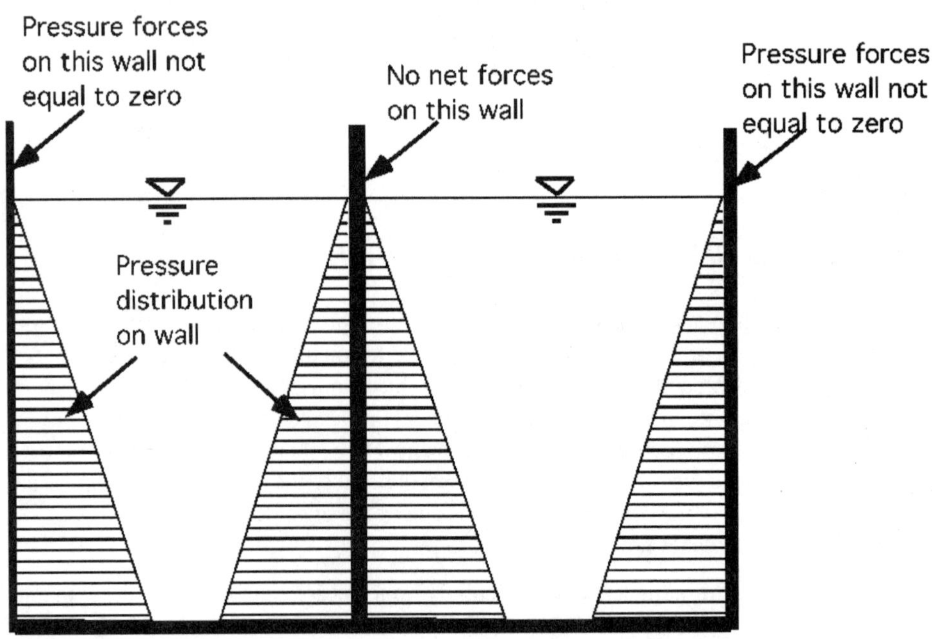

FIGURE 2.5.2 Pressure forces on submerged walls.

2.6 FORCES ON PLANE SUBMERGED SURFACES

P_o = Ambient pressure

γ = Specific weight of fluid

Note: $\bar{y} = \dfrac{\bar{h}}{\sin\theta}$ or $\bar{h} = \bar{y}\sin\theta$

Force = (average pressure) × (area of the surface)

$= (\gamma\bar{h})(A)$

The location of center of pressure, y_p (as shown) is given by

$$y_p = \bar{y} + \frac{I_{xx}}{\bar{y}A}$$

where I_{xx} = Moment of inertia about the centroidal axis of the surface.

In general, therefore, y_p is greater than $\bar{y}$ and the center of pressure is therefore a little below the centroidal ($\bar{y}$) of the surface.

The ambient pressure usually is ignored in these problems because it acts on both sides. However, there are situations where ambient pressure must be considered in the analysis of forces.

Determine the force on the stopper at B due to the water on the other side. The gate is 3 m wide (into the paper).

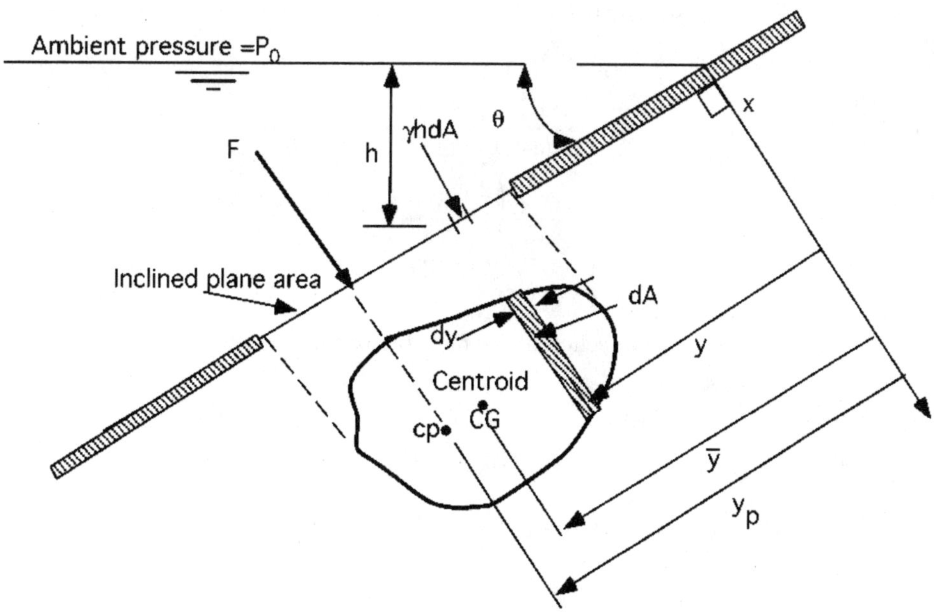

FIGURE 2.6.1 Definition sketch for forces on a plane surface.

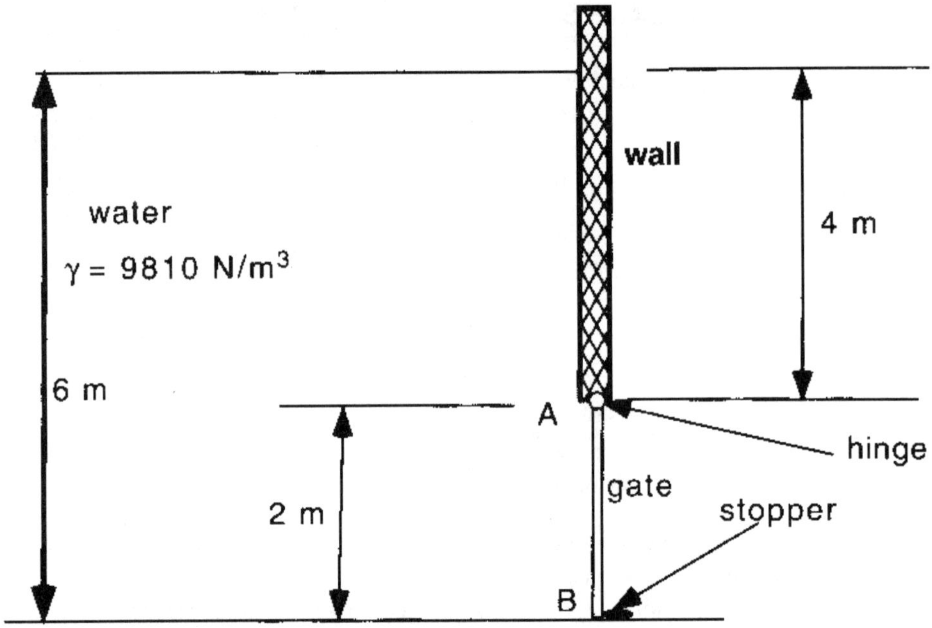

FIGURE E2.6.1a Problem definition sketch for Example 2.6.1.

Solution:

Since the gate is vertical, $\theta = 90°$.

1. Force on the gate due to water,

$$Fw = (\lambda \bar{h})(Area)$$

$$= \left[9810\ \frac{\text{N}}{\text{m}^3}\right][5\ \text{m}][2 \times 3\ \text{m}^2]$$

Note that $\bar{h}$ = Distance to the centoid of the gate

$$= 4\ \text{m} + \left(\frac{2}{2}\right)\text{m} = 5\ \text{m}$$

$$\therefore\ Fw = [9810 \times 5 + 6] = 294{,}300\ \text{N}$$

$$= 294.3\ \text{kN}$$

2. The line of action of Fw must be found.

$$y_p = \bar{y}\ \frac{I_{xx}}{\bar{y}A}$$

But here, since $\theta = 90°$, $\bar{y} = \bar{h}$ *and* $y_p = h_p$

$$I_{xx} = \frac{1}{12}bh\ \text{(Refer to Appendix)} = \frac{1}{12}(3 \times 2^3) = 2$$

$$\therefore\ Y_p = h_p = 5 + \frac{2}{5 \times 6} = 5.0667\ \text{m}$$

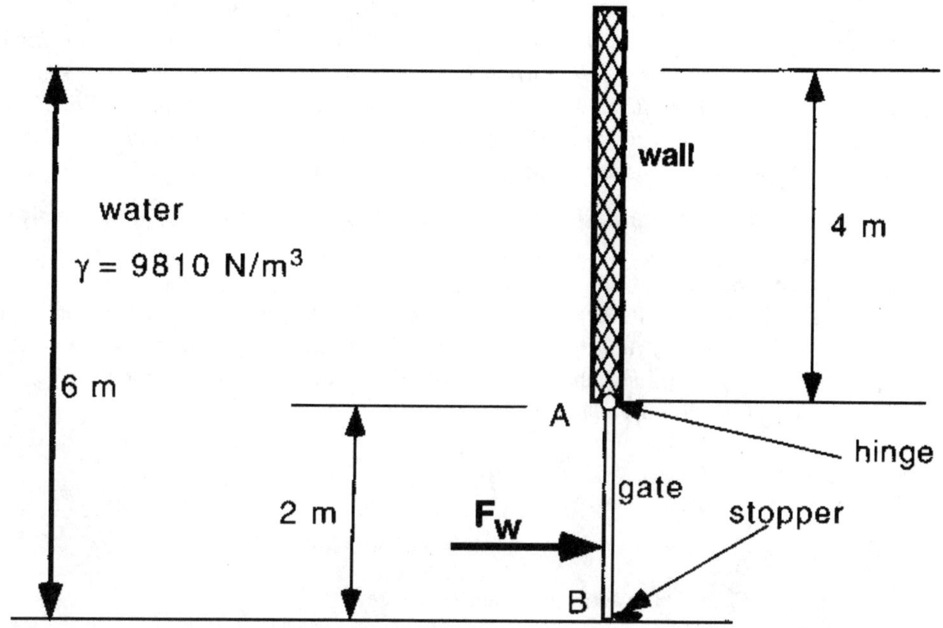

FIGURE E2.6.1b Solution sketch for Example 2.6.1.

3. The free body diagram is shown below

Taking moments about the hinge:

$$\Sigma m = 0 = Fw(1.0667) + F_s(4) = 0$$

$$\therefore\ Fs = \frac{1.0667Fw}{4} = \frac{(1.0667)(294.3)}{4}$$

$$= 74.07 \text{ kN}$$

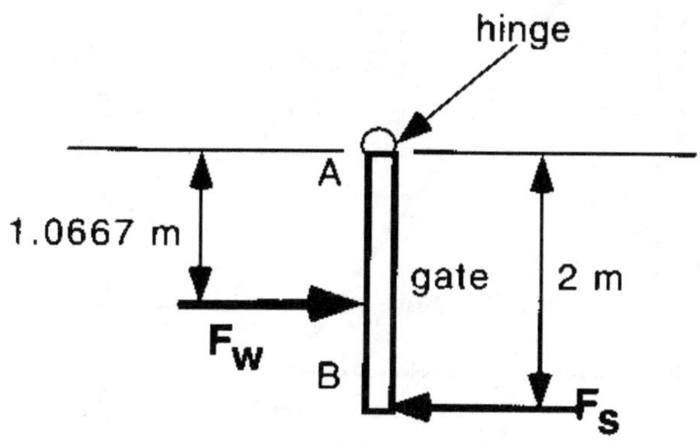

FIGURE E2.6.1c Free body diagram for gate in Example 2.6.1.

EXAMPLE 2.6.2. FORCES ON A SIMPLE DAM CROSS SECTION

For the dam shown, a new experimental medium-weight concrete is being considered (S = 1.8). The depth of water behind the dam ($\gamma = 9810$ N/m³) h = 90 m. The dam is (L = 100 m) high, W = 90 m, D = 20 m and tailwater depth t = 6 m. Determine the magnitude of the forces necessary to properly analyze the stability of the dam. Will the dam topple? Assume that uplift pressure is hydrostatic.

Note: In this analysis, only a unit length of the dam cross-section is analyzed.

Solution:

The forces on the dam are as shown on the figure. These forces are F_h (due to the water behind the dam), F_u (uplift force under the dam), F_t (tailwater force on the downstream face of the dam) and F_{w1} and F_{w2} (weight of the dam—to two components).

Force on upstream wall of dam:

$$F_u = \gamma \bar{h} A = (9810)\left(\frac{90}{2}\right)(90 \times 1)$$
$$= 39{,}730.5 \text{ kN}$$

F_h acts 60m from the water's surface.

Uplift Forces

The uplift force = $F_{u1} + F_{u2}$. The uplift force has been broken into two components to speed up the solution process and because the line of action of each is known.

$$F_{u1} = \gamma(t)(\text{Area}) = 9810(6)(90 \times 1)$$
$$= 5297.40 \text{ kN acting at 45m from A.}$$

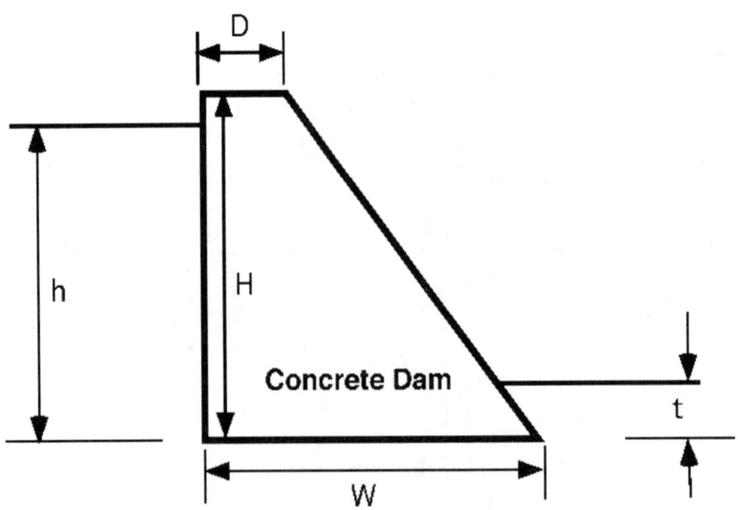

FIGURE E2.6.2a Dam cross section for Example 2.6.2.

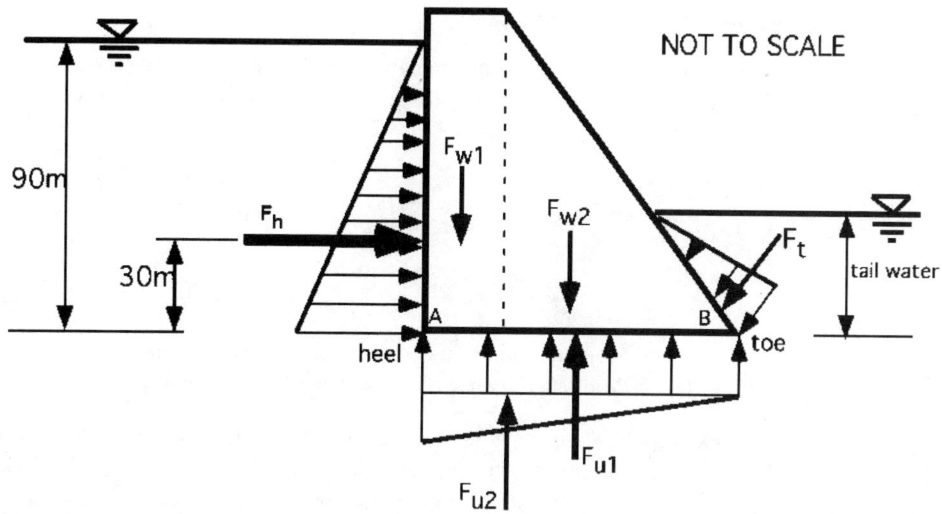

FIGURE E2.6.2b Forces on a dam.

$$F_{u2} = \gamma\left(\frac{90 - t}{2}\right)(\text{Area}) = 9810\left(\frac{90 - 6}{2}\right)(90 \times 1)$$

$$= 37{,}081.8 \text{ kN acting 30m from A.}$$

Tailwater Force

$$\theta = \tan^{-1}\left(\frac{100}{70}\right) = 55.0°$$

$$K = \frac{6}{\sin\theta} = \frac{6}{0.8192} = 7.32 \text{ m}$$

$$F_t = \gamma\left(\frac{t}{2}\right)(7.32 \times 1) = (9810)\left(\frac{6}{2}\right)(7.31 \times 1) = 215.54 \text{ kN}$$

This force acts $\frac{7.32}{3}$ m = 2.44 m from B on the force of dam.

If the ambient pressure (Pa) must be considered in the analysis, then the following issues must be addressed in all calculations:

1. Force on the plane surface:

$$\text{Force} = (\text{Pa} + \gamma\bar{h})\text{A}$$

2. Location of the center of pressure:

$$y_p = \bar{y} + \frac{\gamma \sin\theta\, I_{xx}}{(\text{Pa} + \gamma\bar{h})A}$$

Note that the center of pressure depends on γ when the ambient pressure Pa is considered.

Forces on Asymmetrical Plane Surfaces

In most cases, the plane surfaces considered are symmetrical, but when a surface is not symmetrical, the x component of the force must be considered, since it is on the x-component of the centroid.

$$I_{xx} = \frac{1}{12}\,bh^2 \qquad I_{xx} = \frac{1}{36}\,bh^3$$

$$I_{xy} = 0 \qquad I_{xy} = \frac{b(b-2a)h^2}{72}$$

where I_{xx} = Moment of inertia about the centroidal axis in the plane of the surface
I_{xy} = Product of inertia of the surface (zero for symmetrical geometries).

When the surface is asymmetrical, the resultant force does not lie on the centroidal axis (in the lateral direction of the surface). This can also result in a significant lateral moment, which can be important depending on the particular situation.

The location of the center of pressure in the lateral direction (x-direction) from the centroid is given by:

$$x_p = \frac{-\gamma \sin\theta\, I_{xy}}{(\text{Pa} + \gamma\bar{h})A}$$

In order to accurately interpret this, with reference to Figure 2.6.3, obviously I_{xy} can be 0, positive or negative. When I_{xy} is negative, x_p is positive, and vice versa.

To reiterate, when I_{xy} is zero (symmetrical plane), the point p lies on the $y-y$ centroidal axis below the centroid 0.

Ambient Pressure and Hydrostatic Forces on Surfaces

In most situations, the ambient pressure (usually the atmospheric pressure) is not considered in calculations because the ambient pressure acts on both sides of a surface. For example, in calculating the hydrostatic forces on the side of a ship, it is recognized that the atmospheric

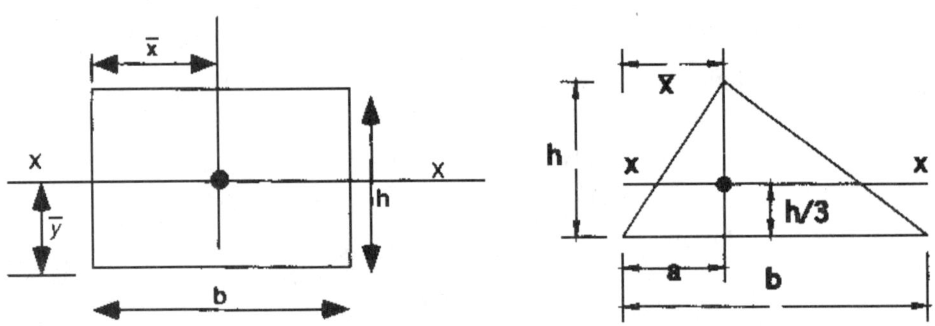

FIGURE 2.6.2 Example of symmetrical (rectangular) and an asymmetrical (triangular) plane surface.

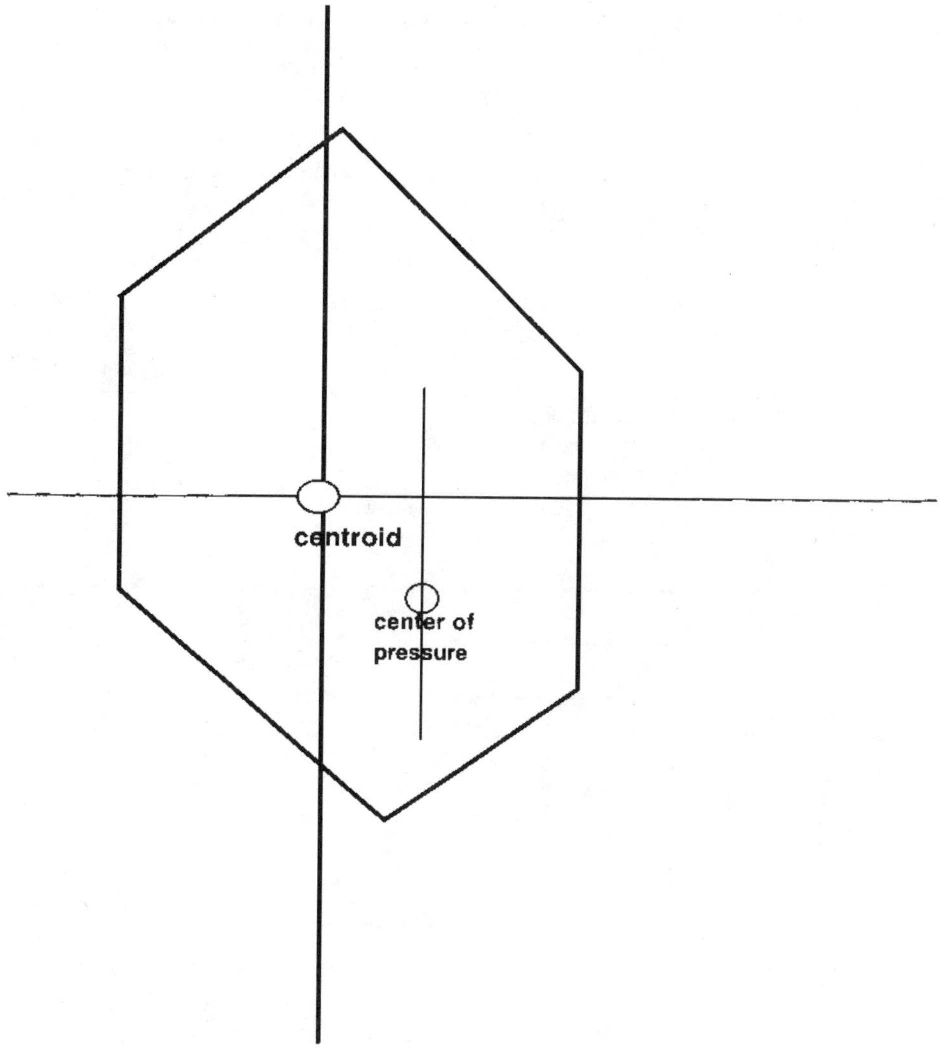

FIGURE 2.6.3 Definition sketch showing the center of pressure, the centroid on an asymmetrical plane surface.

pressure on both sides of the ship's hull would result in net zero force. Similar considerations are implied when analyzing a gate, or forces on a dam.

EXAMPLE 2.6.3

1. *Determine the force on the triangular section of a tank containing glycerine under pressure, $P = 40$ kPa gage.*

   With reference to Figure E2.6.3.1, $H = 4$ m, $a = 6$ m, $b = 3$ m and $\theta = 25°$.

2. *Calculate the location of the center of pressure in the longitudinal (y) and lateral (x) directions.*

TABLE 2.6.1. Summary Equations for Forces on Plane Surfaces

	When ambient pressure (Pa) is considered	When ambient pressure (Pa) is *not* considered
Resultant force F	$= (\text{Pa} + \gamma\bar{h})A$ $= (\text{Pa} + \gamma\bar{y}\sin\theta)A$	$= \gamma\bar{h}A$ $= \gamma\bar{y}\sin\theta A$
Slant (longitudinal) distance to center of pressure, y_p	$= \bar{y} + \dfrac{\gamma \sin\theta\, I_{xx}}{(Pa + \gamma\bar{h})A}$ Since y_p depends on γ it would be different for different fluids	$= \bar{y} + \dfrac{I_{xx}}{\bar{y}A}$ Note that y_p is only dependent on the geometry of the surface
Slant (lateral) distance of COP from yy centroidal axis	$y_p = \dfrac{-\gamma \sin\theta f_{xy}}{(Pa + \gamma\bar{h})A}$ Notice that here x_p depends on γ and Pa as above.	$= -\dfrac{I_{xy}}{\bar{y}A}$

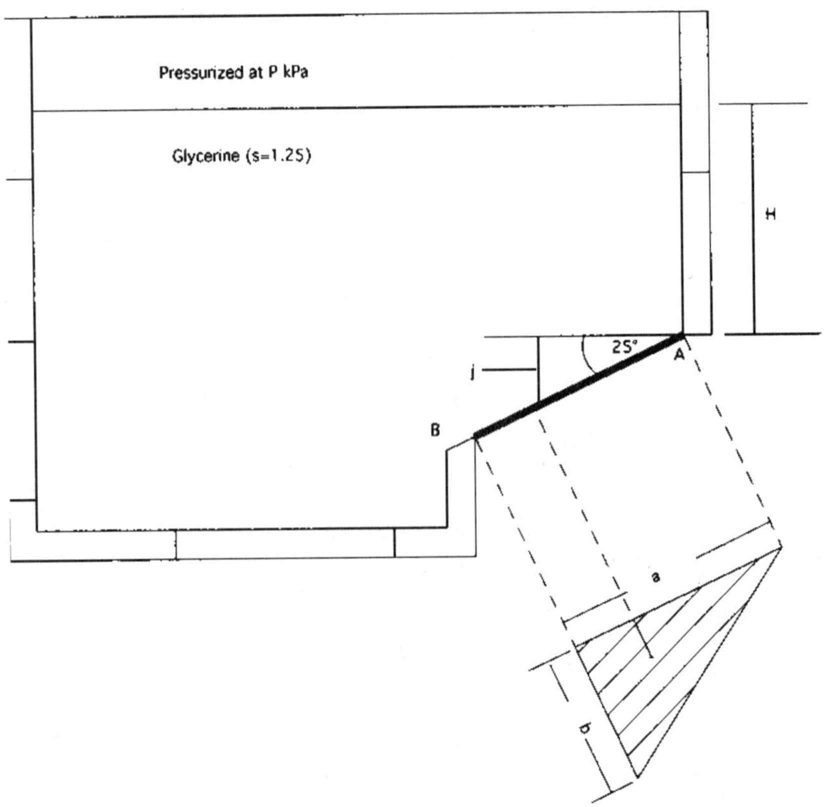

FIGURE E2.6.3a Definition sketch for Example 2.6.3.

Solution:

In this problem, a situation arises where the ambient pressure and the asymmetry of the surface in question must be considered.

The force on the triangular section of tank must take the pressure above atmospheric pressure (Pa = 40 kPa) into account.

$$\bar{h} = H + j$$

where $j = \left(\frac{2}{3}a\right)\sin\theta$

Note that the slanted distance (along the triangular section from A to the centroid of the triangular section $= \frac{2}{3}a$.

$$\therefore \bar{h} = 4 + \frac{2}{3}(6)\sin 25° = 5.6905 \text{ m}$$

1. Force $= (\text{Pa} + \gamma\bar{h})A = [40 \times 10^3 + (1.25)(9810)(5.6905)]x\left(\frac{1}{2} \times 3 \times 6\right)$

$= 988.0$ kN

2. To determine the (x,y) coordinate of the center of pressure:

$$y_p = \bar{y} + \frac{\gamma \sin\theta\, I_{xx}}{(\text{Pa} + \gamma\bar{h})A}$$

$$\bar{y} = \frac{\bar{h}}{\sin\theta} = \frac{5.6905}{\sin 25°} = 13.465 \text{ m}$$

$$\therefore y_p - \bar{y} = \frac{\gamma \sin\theta\, I_{xx}}{(\text{Pa} + \gamma\bar{h})A} \quad \text{(distance from the centroid)}$$

$$= \frac{1.25(9810)(\sin 25°)I_{xx}}{[40 \times 10^3 + (1.25)(9810)(5.6905)]} = 0.04720\, I_{xx}$$

But

$$I_{xx} = \frac{1}{36}ba^3 = \frac{1 \times 3(6)^3}{36} = 18\text{m}^4$$

$$\therefore y_p - \bar{y} = 0.047207 \times 18 \text{ m}^4 = 0.850 \text{ m}$$

The x-coordinate of the center of pressure is found similarly but by using the product of inertia, I_{xy}.

The distance of the lateral location of the center of pressure from the centroid of the plane surface is found as follows:

$$x_p = -\frac{-\gamma \sin\theta I_{xy}}{(\text{Pa} + \gamma\bar{h})A}$$

$$I_{xy} = \frac{b(b - 2a)h^2}{72}$$

$$\therefore I_{xy} = \frac{b(b - 2a)h^2}{72}$$

Here, $a = 3$ m:

$$\therefore I_{xy} = \frac{3(3-6)6^2}{72} = -4.5 \text{ m}^4$$

$$x_p = \frac{-1.25(9810)(\sin 25°)(-4.5)}{\{40{,}000 + [(1.25)(9810)(5.6905)]\}}$$

Geometrically, this will look like Figure E2.6.3c below.

2.7 FORCES OF CURVED SURFACES

In order to address hydrostatic forces on curved surfaces, it is very effective to separate the forces into vertical and horizontal components, calculate their respective magnitudes, and then combine the forces to obtain the resultant (total) force on the curved surfaces.

Consider the following arbitrary curved surface that is submerged in a certain fluid as shown in Figure 2.7.1.

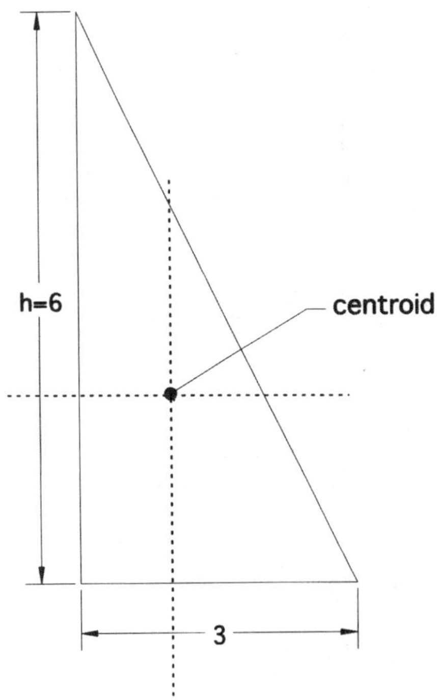

FIGURE E2.6.3b Location of centroid on a triangular surface.

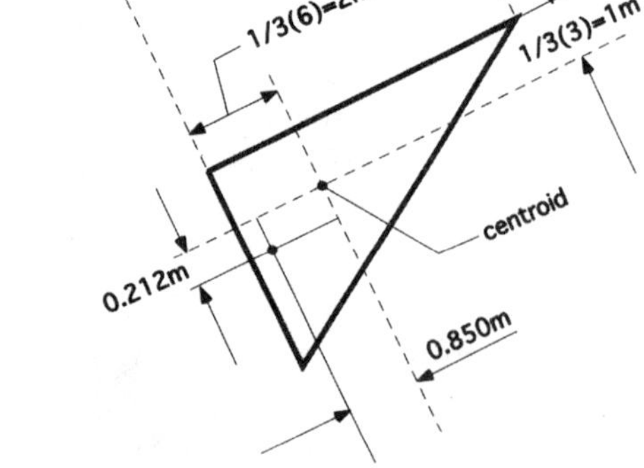

FIGURE E2.6.3c Schematic presentation of orientation of forces.

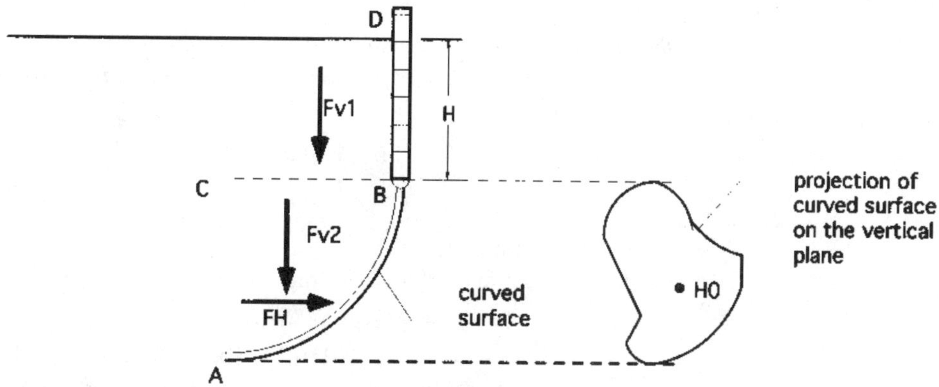

FIGURE 2.7.1 Definition diagram for forces on a curved surface.

In Figure 2.7.1, F_v = vertical hydrostatic force on the curved surface. F_v is equal to the *sum* of force (weight) of the fluid directly *above* the curved section (F_{v2}) and the weight of the fluid above CB. The horizontal hydrostatic force is equal to the force on the projected vertical surface as shown. Here H_o is the center of pressure of the *projected surface.*

The magnitude of the hydrostatic force, F_R is given by:

$$\sqrt{F_{H^2} + F_v^2}$$

The line of action of F_R is easily determined by the fact that its orientation is such that

$$\alpha = \tan^{-1}\left(\frac{F_V}{F_H}\right)$$

Note that the ambient pressure might need to be considered in certain situations. One example would be a case where the surface of fluid is under a gage pressure other than zero (positive or negative gage pressure).

The application of these principles is further illustrated in the following example.

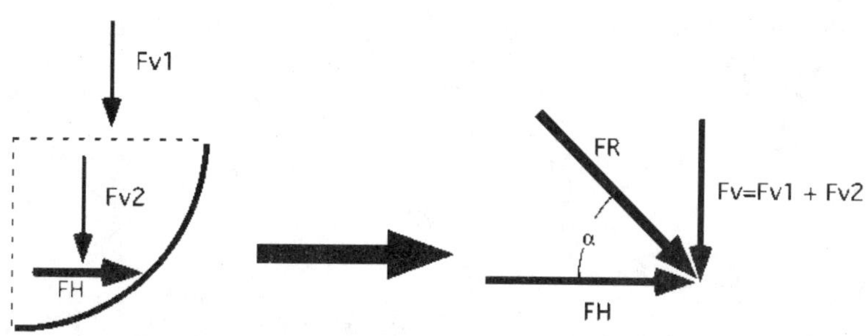

FIGURE 2.7.2 Hydrostatic force components and resultant on a curved surface.

EXAMPLE 2.7.1

Calculate the hydrostatic force components and the resultant force on the curved surface given in Figure E2.7.1.1. The tank is pressurized since it contains a carbonated beverage at 15 kPa_{gage} with a density,

$$\gamma = 900 \text{ kg/m}^3$$

Solution:

1. Vertical hydrostatic forces:
   a. Force due to pressure on top of beverage

$$F_{vp} = \text{(pressure)(Area)}$$
$$= (15 \times 103)(2 \times 10) = 300 \text{ kN}$$

   b. Force due to weight of carbonated beverage in CBDE = (Volume)(Specific weight)

$$F_{v1} = (3 \times 2 \times 10)(900 \times 9.81) = 529.74 \text{ kN}$$

   c. Force due to weight of fluid in ABC:

$$F_{v2} = \forall_{ABC} \times \gamma = \left[\left(\frac{\pi 2^2}{4}\right)10\right][900 \times 9.81] = 277.37 \text{ kN}$$

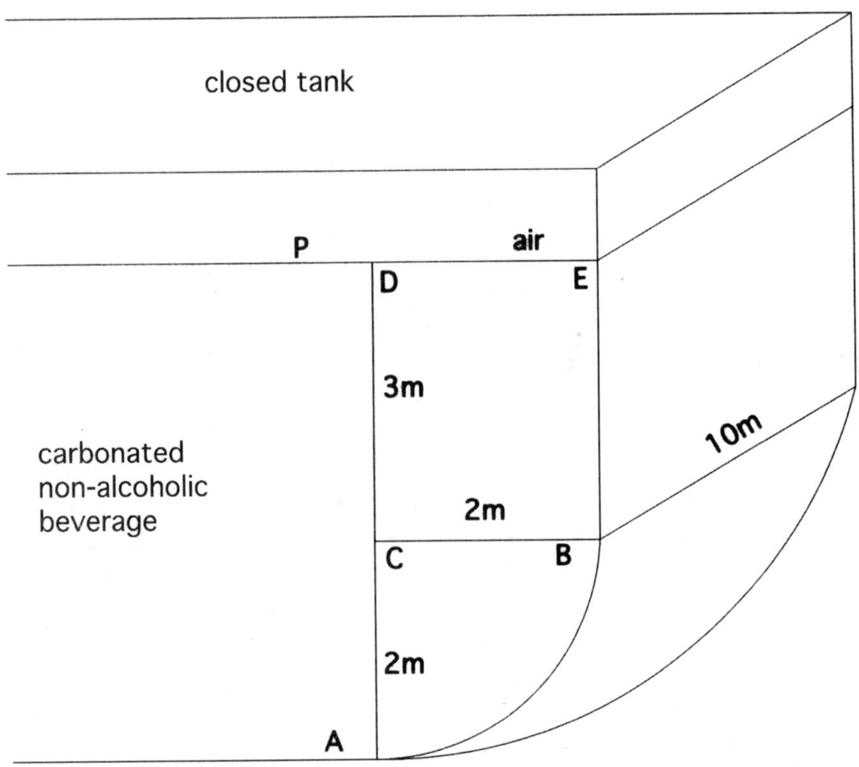

FIGURE E2.7.1a Diagram for Example 2.7.1.

2. Horizontal components:

$$F = (P + \gamma \bar{h})A$$

Here P = 15 kPa = 15,000 $\frac{\text{N}}{\text{m}^3}$

$$\bar{h} = \left(3 + \frac{2}{2}\right) = 4 \text{ m}$$

$$\therefore F_H = [15{,}000 + (900)(9.81)](2 \times 10)$$
$$= 476.58 \text{ kN}$$

3. The total or resultant force, F_R is given by

$$F_R = \sqrt{F_{V^2} + F_{H^2}}$$

$$F_V = F_{VP} + F_{V1} + F_{V2} = 300 + 529.74 + 277.37$$
$$= 1106.74 \text{ kN}$$

$$\therefore F_R = \sqrt{1106.74^2 + 476.58}$$
$$= 1205.0 \text{ kN}$$

4. The line of action of the force:

$$F_v = 1106.74$$

$$\alpha = \tan^{-1}\left(\frac{F_V}{F_H}\right)$$

$$\alpha = \tan^{-1}\left(\frac{1106.74}{476.58}\right) = 66.70°$$

2.8 ARCHIMEDES' PRINCIPLE

Archimedes' Principle can be summarized thus:

1. A body wholly or partially immersed in a fluid will be *buoyed up* by a *force* (experience an upward force) equal to the weight of the fluid it displaces.
2. The resultant buoyant force acts upward and at the center of mass of the displaced fluid and *not* necessarily at the center of mass of the floating or immersed body. The location of the resultant buoyant force is called the *center of buoyancy*.

EXAMPLE 2.8.1

A metal alloy anchor appears 405 N lighter in the Atlantic Ocean than in air.

a. What is the volume of the anchor?
b. How much does it weigh in air and in the ocean? The density of the anchor metal alloy is 9000 kg/m^3 and the specific gravity of the ocean water is 1.05.

Solution:

a. Since the anchor weighs 405 N lighter in the ocean, the buoyant force acting on it = 405 N (according to Archimedes' principle).

Further, by Archimedes' principle, an immersed body *displaces its volume* and the buoyant force experienced by it is equal to the *weight of the fluid* it has displaced. Therefore,

Volume of the metal alloy anchor = Volume of ocean water that weighs 405 N.

$$\text{Specific weight of ocean water, } \gamma = 1.05 \times 9810 \frac{\text{N}}{\text{m}^3} = 10{,}300.5 \frac{\text{N}}{\text{m}^3}$$

$$\text{Volume of anchor} = \frac{405 \text{ N}}{10{,}300.5 \dfrac{\text{N}}{\text{m}^3}} = 3.931 \times 10^{-2} \text{ m}^3$$

b. Weight (in air) = Mass × g = Volume × Density × g

$$\text{Weight of anchor} = 3.931 \times 10^{-2} \text{ m}^3 \times 9000 \frac{\text{kg}}{\text{m}^3} \times 9.81 \text{ m/s}^2 = 3471.4 \text{ N}$$

$$\begin{aligned}\text{Weight of anchor in water} &= \text{Weight (air)} - \text{buoyant force} \\ &= (3471.4 - 405.0) = 3066.4 \text{ N}\end{aligned}$$

EXAMPLE 2.8.2

A barge weighs 55 metric tons when empty (without any cargo). The barge is 10 m wide, 15 m long and 3 m high. Determine the portion of the barge that is below the water line when 125 metric tons of military hardware are loaded onto it and it is navigating on the Atlantic Ocean (specific gravity = 1.10).

(Note: 1 metric ton = 1000 kg_f = 2000 lb; 1 kg_f = 9.8067 N)

Solution:

$$\begin{aligned}\text{Total weight of barge with cargo} &= 55 + 125 = 180 \text{ metric tons} \\ &= 180 \times 1000 \times 9.8067 \text{ N} \\ &= 1.765 \times 10^6 \text{ N}\end{aligned}$$

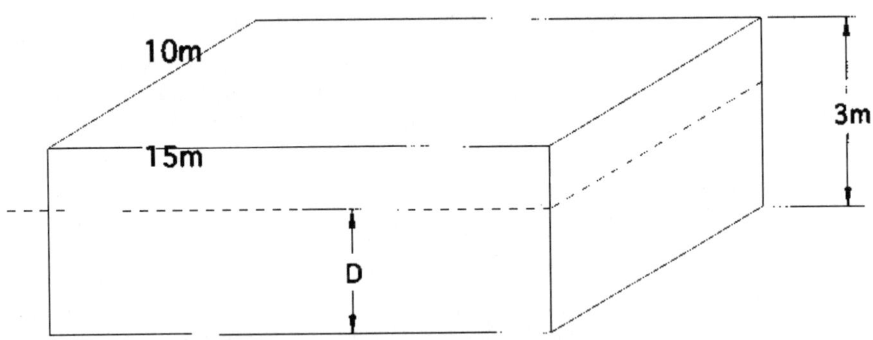

FIGURE E2.8.2a Definition sketch for Example 2.8.2.

The barge must displace its weight to float.

$$\therefore \text{Volume displaced} = \frac{\text{Total weight of barge}}{\text{Specific weight of ocean water}}$$

$$= \frac{1.765 \times 10^6 \text{ N}}{(1.10)9810 \dfrac{\text{N}}{\text{m}^3}} = 163.58 \text{ m}^3$$

The portion of the barge below the water line, D:

$$D = \frac{\text{Volume displaced by barge}}{\text{Cross-sectional area of barge}} = \frac{163.58 \text{ m}^3}{10 \times 15 \text{ m}^3}$$

$$\therefore D = 1.09 \text{ m}$$

CHAPTER 3
FLUID DYNAMICS AND APPLIED HYDRAULICS

In this chapter, basic concepts in fluid dynamics are presented. Further, practical issues in applied hydraulics are examined. In fact, the chapter can be conveniently divided into a theory section and an application section. The flow of fluids (gases and liquids) in pressurized and free-surface conduits (open channel flow), but particularly water, is presented. In this regard, the flow of water in pressurized conduits is examined in detail. The characterization of fluid flow as 1-dimensional, 2-dimensional or 3-dimensional is discussed. The mathematical representations of a fluid field by the Lagrangian or Eulerian viewpoints is presented. Regimes of flow, as determined by the value of the Reynolds number, which categorizes fluid flow as laminar, in transition, or turbulent, are covered. The chapter also covers the following topics and concepts: the Reynolds Transport Theorem, Conservation Laws (conservation of mass, conservation of energy, and conservation of linear momentum), velocity distribution in pressurized conduits and open channels, the calculation and use of momentum and energy flux correction factors, friction headloss and the Darcy-Weisbach equation, minor headlosses, pumps and pipe networks (including the Hardy Cross method for network analysis), flow in open channels and the design of rigid- and movable-boundary channels.

3.0 FLUIDS IN MOTION

This section briefly presents the basic fluid flow concepts that provide the necessary foundation to clearly understand the fluid dynamics methods described in the sections that follow. The material presented here includes:

- The characterization of fluid flow regimes
- The concept of 1-, 2-, and 3-dimensional flows
- Flow regimes (laminar, transitional, and turbulent)

- The idea of the viscous sublayer
- Techniques for measuring flowrate in pipes (pressurized conduits) and open channels
- The Bernoulli equation
- The Reynolds Transport Theorem and associated application to conservation laws

The primary focus of this part of the book is to set forth the underlying principles that provide a framework for the techniques used in solving fluid flow problems the reader might encounter in the practice of civil engineering.

3.0.1 Characterization of Fluid Flows

The following topics are presented in this section:

1. Categorization of fluid flows as 1-, 2-, and 3-dimensional
2. Eulerian and Lagrangian viewpoints of fluid flow
3. Steady and unsteady flow
4. Uniform and non-uniform flow
5. Relative motion
6. Pathlines, streaklines, and streamlines

Categorization of Fluid Flows as 1-, 2-, and 3-dimensional. Generally, all fluid flow can be categorized as 3-dimensional and can be described in Cartesian coordinates as having component values in the three orthogonal directions (x,y,z). Further, most flows generally are time-dependent, and thus the velocity at a particular point can be described as $\mathbf{V}(x,y,z,t)$, where $\mathbf{V}$ is a vector quantity.

When certain conditions prevail, however, a flow field can be described adequately with less than the usual three dimensions. The mathematics (effort and complexity) of dealing with a three-dimensional problem usually is reduced significantly if one can assume that the dimensional space is less than three dimensions. Therefore, whenever justifiable, assumptions are made to reduce the number of dimensions to be used in describing a flow. In general, the simplification is allowable (in fact, sought) whenever it does not result in a significant misrepresentation of the realities of the actual flow field. A few example situations where simplification is possible follow.

Examples of 1-dimensional flow:

1. Steady flow in a long, straight pipe (with constant diameter)
2. Steady fluid flow in a long, open channel with a constant cross-section laid at a constant longitudinal slope (a prismatic channel)
3. Steady fluid flow in a constant diameter conduit with some degree of curvature

Two-dimensional fluid flow can be described as flows in which the velocity vector field is practically constant or negligible in the third dimension but significantly variable or constant in the remaining two dimensions.

Examples of 2-dimensional flows are:

1. Flow over an airfoil
2. Flow over a long, broad-crested weir (away from the lateral edges of the nappe)
3. Flow over the crest of an overflow spillway of a long (laterally) dam
4. Flow around a cylinder

Lagrangian and Eulerian viewpoints. In the previous section, it was noted that the velocity field in a fluid flow can be described using a multivariate function of the position in Cartesian coordinates and time. Therefore, the velocity field can be written as $V = v(x,y,z,t)$.

By the Lagrangian viewpoint, the flow field is described mathematically by focusing on a particle or a group of particles in the flow field. For these fluid elements or particles, the position, velocity, and temperature might be measured or calculated. In the Lagrangian description of fluid flow, the important distinction is that particles or fluid elements are tracked or monitored in order to describe the flow field mathematically. Hence, for particles in the flow field at a particular time, t:

$$\mathbf{r}(t) = x\mathbf{i} + y\mathbf{j} + z\mathbf{k}$$

where $\mathbf{r}(t)$ = Position vector at time t
$\mathbf{i}, \mathbf{j}, \mathbf{k}$ = Unit vectors in the 3 orthogonal directions.

and the velocity vector is given by $\mathbf{v}(t) = \dfrac{d\mathbf{r}(t)}{dt}$

and the acceleration, $\mathbf{a}(t) = \dfrac{d\mathbf{V}(t)}{dt}$

The task of tracking particles in a flow can easily become overwhelming when a significant number of particles must be tracked to adequately describe the flow field. This limitation can be addressed with the Eulerian approach.

In the Eulerian viewpoint, the focus is on points in the flow field. The flow field is characterized by determining the values or changes in parameters of interest with respect to time. The flow field is described thus: $V = v(x,y,z,t)$ in Cartesian coordinates, where v is the velocity field. Acceleration will be represented as:

$$a = \frac{dV}{dt} = \frac{dv(x,y,z,t)}{dt}$$

It is usually mathematically less cumbersome to describe a flow field by the Eulerian approach, therefore it is used in the subsequent sections of this book.

Steady and Unsteady Flow. Flow is said (or classified) to be steady if, at a given point in space (x,y,z), flow properties do not change with time. In other words, the flow field is time invariant with respect to the flow parameters that are of interest. Mathematically, for the velocity field, steady flow is defined as:

$$\frac{\partial V}{\partial t} = 0 \quad \text{(flow is steady)}$$

The important idea here is that fluid properties do not vary with time at the particular point of interest. For example, say the flow under a bridge is being monitored and, in particular, the velocity is observed to be unchanging at a certain location, then we can conclude that the flow is steady.

Note finally that if

$$\frac{\partial V}{\partial t} \neq 0,$$

flow is not steady.

Uniform and Non-Uniform Flow. A flow is described as being *uniform* if the properties that are used to describe it do not change with position (or location). The flow does not have to necessarily be steady to be uniform. Mathematically, steady flow is described by

$$\frac{\partial V}{\partial s} = 0.$$

When the preceding is not true, then the flow is *non-uniform*.

Relative Motion (Invariance to the Galilean Transformation). The relative motion of a body in a fluid is the same whether the body moves at constant speed through a stagnant ocean of fluid or whether the fluid streams past a stationary body at constant speed. Fluid mechanics draws no distinction between the two cases. The results of each case can be interpreted without reference to which experimental method is used. This is called *invariance to the Galilean transformation*. It is sometimes easier to both watch and understand the mechanics of a fluid passing a stationary body, whereas the practical result desired is more frequently associated with a body moving through a still fluid. Examples include wind tunnel studies of airfoils, buildings, automobile shapes, and studies of submarine or torpedo shapes in water.

Pathlines, Streaklines, and Streamlines in a Flow Field. In the description of a flow field, three types of lines have been used to effectively provide ways to visualize fluid flow. The *pathline* is the actual path traversed by a given fluid particle or element in a flow field. A pathline is therefore the history of a particle's movement in a flow field. As a physical concept, the pathline can be captured, for example, by a time-exposure photograph of the path of a dyed fluid particle in the flow field.

A *streakline* is the locus of particles, at a particular future time, which have earlier passed through a prescribed point. Streaklines are a physical concept and therefore can be produced experimentally by the continuous release of marked fluid elements (dye, smoke, or bubbles) from a given point. If a flow is steady (properties do not change with time), then pathlines and the streaklines are coincident. This must be so because every particle that passes through a particular upstream point (when flow is steady) shares the same pathline as it moves downstream.

A *streamline* is a mathematical concept, not a physical one. Streamlines are vital for the mathematical description of fluid flow. A streamline is a line that is everywhere tangential to the velocity vector at any instant. The streamline cannot be photographed directly. When flow is steady, however, pathlines, streaklines and streamlines are coincident. It is important to note some of the properties of streamlines:

1. By definition, the velocity vector in the flow is tangent to streamlines everywhere.
2. Streamlines never cross each other. If streamlines could cross, then it would be possible to have more than one velocity value (magnitude and direction) at a particular point (which is impossible).
3. The spacing of streamlines is inversely proportional to the magnitude of velocity in the flow field. Thus, a closer spacing of streamlines implies higher velocity values.
4. If the flowing fluid is bounded by walls, then the walls are also streamlines. (Notice that the no-slip condition is consistent with the concept of streamlines.)
5. When streamlines are parallel and straight, the flow can be described as steady and uniform, and the pressure gradient across the section is zero.
6. When a set of lines are drawn at right angles (90°) to streamlines, they are called *equipotential lines*.

3.0.2 Forces on a Fluid Element

The forces on a fluid element can be attributed to *gravitational acceleration*, *pressure*, *viscosity*, *elasticity*, *surface tension*, and *inertia*. In many fluid flow problems, however, some of these forces are either not present or insignificant. For example, in observing fluid flow past a body immersed in a fluid, the only significant forces involved are those of friction and inertia. Furthermore, rather than be concerned with the exact values of the forces acting on individual fluid particles, an approximate *order-of-magnitude* estimate of the *relative* sizes of these forces might suffice, and indeed is all that is important in certain cases. An example of this is presented below for the determination of the Reynolds Number.

Gravitational forces, which result from the gravitational field, are important in free surface flows (for example, flow in open channels). The Froude Number is a dimensionless number that provides the order of magnitude measure of the ratio of *inertial* forces (see below) to *gravitational* forces. Inertial forces in a fluid result from the acceleration or motion of the fluid particles. Recall that *Force* = *Mass* (*Acceleration*). Viscous forces develop from shear stresses due to the viscosity of the fluid. Energy loss due to friction in a pipe flow is because of the viscosity of the fluid flowing in the pipe. Recall that, shear stress,

$$\tau = \mu \frac{du}{dy}$$

at the wall of the pipe, where μ is the dynamic viscosity and $\frac{du}{dy}$ is the velocity gradient.

3.0.3 Derivation of Reynolds Number

Reynolds Number (Re) represents the ratio of inertial forces to viscous forces,

$$\text{Re} = \frac{\text{Inertial forces}}{\text{Viscous forces}}.$$

It is named after Osborne Reynolds (1842–1912), a British engineer and scientist who discovered the principle while investigating the transition from *laminar* to *turbulent* flow in small tubes. The inertial forces acting on a typical fluid particle are determined by taking the product of its mass and acceleration, as noted previously. The mass per unit volume is, by definition, the density (ρ), and the volume of the fluid element is proportional to the length dimension cubed (L^3). By definition, acceleration is the rate of change of velocity with time during which this change occurs, or in dimensional terms,

$$\frac{L}{T^2}$$

Thus:

$$\text{Inertial Force} = MA = \rho L^3 \frac{L}{T^2} = \frac{\rho L^4}{T^2} = V^2 L^2 \rho$$

where V is the velocity in the flow.

The magnitude of the viscous forces is determined by the product of the viscous shear stress and the surface area (A) over which it acts. The surface area is proportional to the square of a *characteristic length*, or L^2. The viscous stress is proportional to the viscosity (μ) and to the rate of change of speed with distance (velocity gradient or strain rate). Thus:

$$\text{Viscous Force} = \mu \frac{d\mu}{dy} A = \mu \frac{V}{L} L^2 = \mu VL$$

The ratio of inertial forces to viscous forces is

$$\frac{\text{Inertial Forces}}{\text{Viscous Forces}} = \text{Reynolds Number} = \text{N}_\text{R} = Re$$

Reynolds Number is conventionally defined as:

$$Re = \frac{\rho V^2 L^2}{\mu VL} = \frac{\rho VL}{\mu} = \frac{VL}{v},$$

where v is the kinematic viscosity of the fluid.

Reynolds Number is a dimensionless number that is used to classify flows in one of three categories—*laminar*, *transitional*, and *turbulent*. For most applications, when Re $\leq$ 2000, it is said to be in the laminar regime, but for values of Re $\geq$ 3000 (practically 4000), a flow usually is categorized as being turbulent. The velocity profile in a pipe in which the flow is characterized as laminar is distinctly different from the profile that is characterized as turbulent. When Re falls between 2000 and 3000, it is categorized as transitional. In this range, the behavior of the flow is not stable since it transitions between laminar and turbulent flow.

3.1 THE VISCOUS WALL LAYER

All pipe walls, including those made with smooth materials such as plastic, when examined at the microscopic level, will appear rough. Physically rough pipes, however, can be described as hydraulically smooth depending on the thickness of the viscous wall layer δ_{vis}, relative to the average wall roughness height. In hydraulics, the roughness of the physical pipe material is determined by the average wall roughness height, called the *equivalent sand roughness*, ϵ.

When the viscous wall layer is thicker (higher) than the roughness (average roughness height), it submerges the apparent roughness of the wall. The physical roughness of the pipe does not affect the flow appreciably, making it appear to be smooth. In this flow condition, the pipe is characterized as hydraulically smooth.

On the other hand, if the viscous wall layer is thin, such that the roughness of the pipes is higher than the layer ($\epsilon > \delta_{vis}$), then the pipe wall is hydraulically rough. As will be shown later (Section 3.6.4), the *friction factor*, *f*, used to calculate the loss of energy due to friction in a flow, depends on the relative roughness of a pipe, among other factors. The relative roughness of a pipe is a dimensionless ratio of the equivalent sand roughness (ϵ) and the diameter of the pipe (D). Therefore,

$$\text{Relative roughness} = \frac{\epsilon}{D}$$

The concept of roughness is illustrated in Figure 3.1.1.

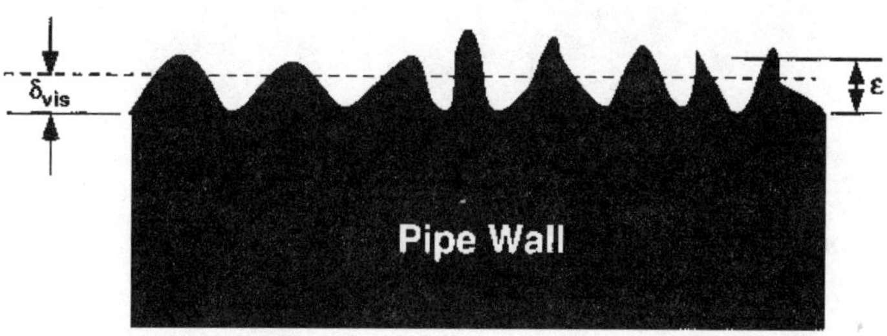

(a) Hydraulically rough pipe flow condition where the thickness of the viscous wall layer, $\delta_{vis} < \varepsilon$ (average height of the conduit's physical roughness).

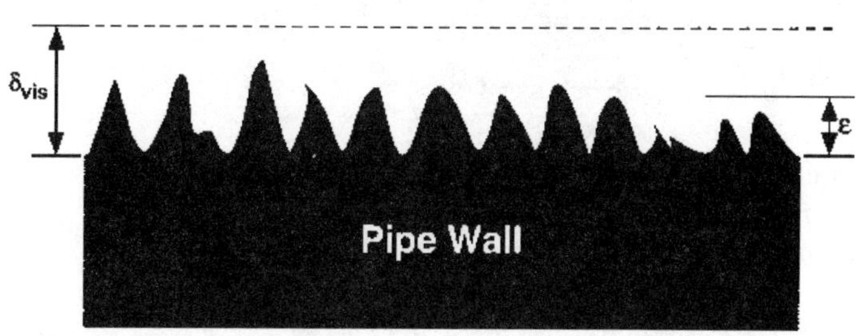

(b) Hydraulically smooth pipe flow condition where the thickness of the viscous wall layer, $\delta_{vis} > \varepsilon$ (average height of the conduit's physical roughness).

FIGURE 3.1.1 Hydraulically rough and smooth conduit flow conditions.

3.2 VELOCITY DISTRIBUTION AND THE MEASUREMENT OF FLOW RATE

The velocity of fluid flowing through a pipe varies from zero (at the pipe wall) to a maximum value at the midpoint of the pipe. The no-slip condition is satisfied because the velocity of flow at the wall is zero, as expected for real (viscous) fluids.

When flow in a conduit is laminar ($Re < 2000$), the velocity distribution is parabolic and the variation of the flow velocity from the wall to the midpoint of the pipe is well-defined. In addition, there is very limited mixing or transfer of momentum from layer to layer, since particles in one layer do not move across other layers. Particles that originate from a common point in one layer tend to trace the same path through the flow field.

When a flow is turbulent (practically, $Re > 4000$) particles tend to trace random paths, and the flow is characterized by mixing. The no-slip condition still is maintained at the pipe wall, but the velocity distribution is not as well defined as in a laminar flow. The difference in velocity of flow from the pipe wall to the midpoint is subdued. Particles in the flow randomly translate from higher velocity points (layers) to lower velocity points, and vice versa, resulting in continuous transfer of momentum from the wall regions to the midpoint sections. Consequently, the resulting equalization of velocity values in the overall flow pattern can be observed. These concepts are depicted in Figure 3.2.2.

When flow in a pipe is laminar, a typical velocity function U_{lam} is given by the following parabolic equation, where $V_{\max}$ is the velocity at the center of the pipe and is the maximum velocity in the flow.

$$U_{\text{lam}} = V_{max}\left(1 - \frac{r^2}{R^2}\right)$$

where $V_{\max}$ = Velocity at the center (midpoint) of the pipe)
R = Radius of the pipe
r = Radial distance from the midpoint (center) of the pipe.

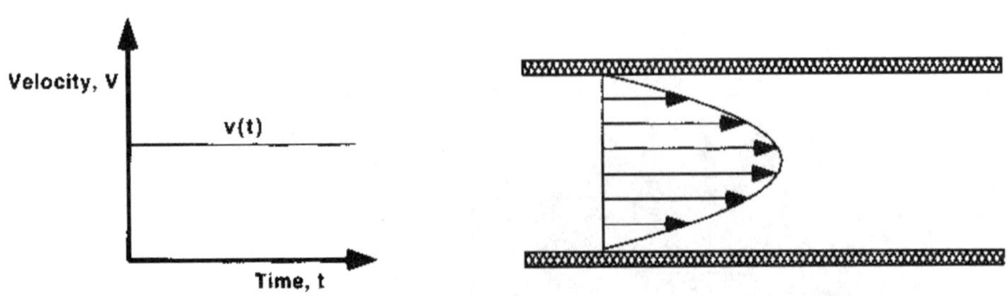

(a) Plot of the velocity of particles passing through a particular point over a period time.

(b) Typical velocity profile in a conduit for laminar flow regime showing a distinct parabolic distribution.

FIGURE 3.2.2 Turbulent flow conditions in a conduit.

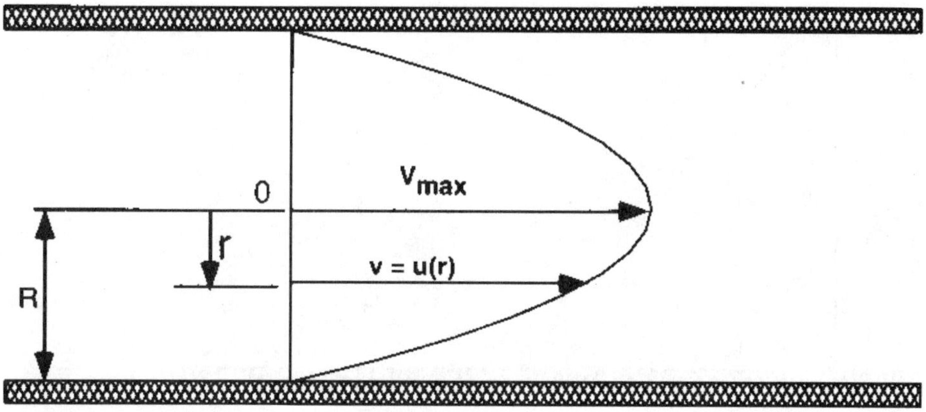

FIGURE 3.2.3 Typical velocity profile for laminar flow.

This is called the *second power law*. Further, it can be shown that the velocity, $V_{\max}$ at the centerline (midpoint) of a pipe (for laminar flow) is given by:

$$V_{\max} = \frac{h_L \gamma}{4\mu L} R^2$$

where h_L = Headloss due to friction in the flow
γ = Specific weight of the fluid
μ = Dynamic viscosity of the fluid
R = Radius of the pipe
L = Length of the pipe

The typical velocity distribution function for turbulent flow in a pipe, U_{turb} is given by:

$$U_{\text{turb}} = V_{\max}\left(1 - \frac{r}{R}\right)^m$$

where $V_{\max}$ = Velocity at the center (midpoint) of the pipe
R = Radius of the pipe
r = Radial distance from the midpoint (center) of the pipe
$m = \frac{1}{7}$

A typical velocity distribution for turbulent flow in a pipe is given by the *one-seventh power law*, which is typical for turbulent flow in piping systems. This is illustrated by Figure 3.2.4.

Flow Rate and Velocity Measurement in Pipes

The volumetric flow rate and average velocity of flow can be determined by:

1. Direct measurement using flow meters

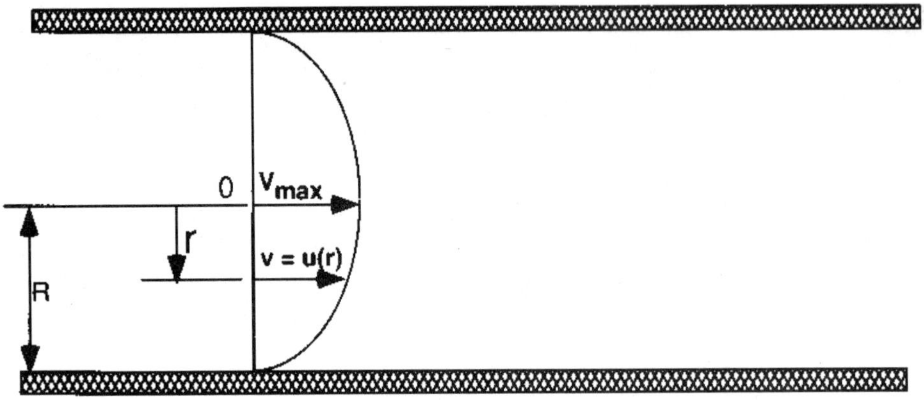

FIGURE 3.2.4 Typical velocity profile for turbulent flow.

2. Indirect measurement by relating the pressure drop across an inline orificemeter, nozzle meter or venturimeter to the velocity and flowrate
3. Numerical or analytical integration technique

Direct Flowrate/Velocity Measurement A properly calibrated flowmeter can be used to determine the volumetric flow rate, Q directly. The average flow velocity, $\bar{V}$, can be obtained by dividing the flow rate by the actual area of flow in the pipe,

$$\bar{V} = \frac{Q}{A}$$

where A = Actual area of flow in the pipe.

A variation of this procedure involves determining the temperature and weight or mass of the fluid (usually a liquid) collected in a container over a specified period of time. The density of the liquid is determined for the measured temperature (usually from a temperature-density reference table).

1. Measure the mass, M, collected in a container over time elapsed, Δt.

   The volume of fluid collected, $\text{Vol} = \frac{M}{\rho} = \forall$

   The volumetric flow rate, $Q = \frac{\forall}{T} = \frac{M}{\rho t}$

   If the volume, however, is measured (directly) over time period, Δt, $Q = \frac{\forall}{T}$

2. Determine weight of fluid (liquid), W collected over time period, Δt

$$Q = \frac{W}{\gamma T}$$

Relating Pressure Drop Across a Constriction to Velocity and Flowrate A measured change (drop) in piezometric head (pressure head +position) across a constriction (venturi, orifice, or nozzle) is equated to the velocity head change, h which is proportional to the second power of the velocity in the upstream section of the pipe, V_1.

By definition (assuming no headlosses):

$$h = \left(\frac{P_1}{\gamma} + z_i\right) - \left(\frac{P_2}{\gamma} + z_2\right)$$

$$\therefore v_1^2 = \sqrt{\frac{2gh}{1 - \left(\frac{A_2}{A_1}\right)^2}}$$

Finally, the flowrate,

$$Q = A_1 V_1$$

Numerical or Analytical Integration

For Pipes Calculus or analytical geometry can be used in conjunction with an assumed (or known) velocity distribution profile for the flow. For example, if laminar flow is assumed, the velocity profile,

$$U = U_{\text{max}}\left[1 - \frac{r^2}{R^2}\right]$$

may be appropriate. Here, r is the generic radial distance from the center of the pipe and R is the radius of the pipe.

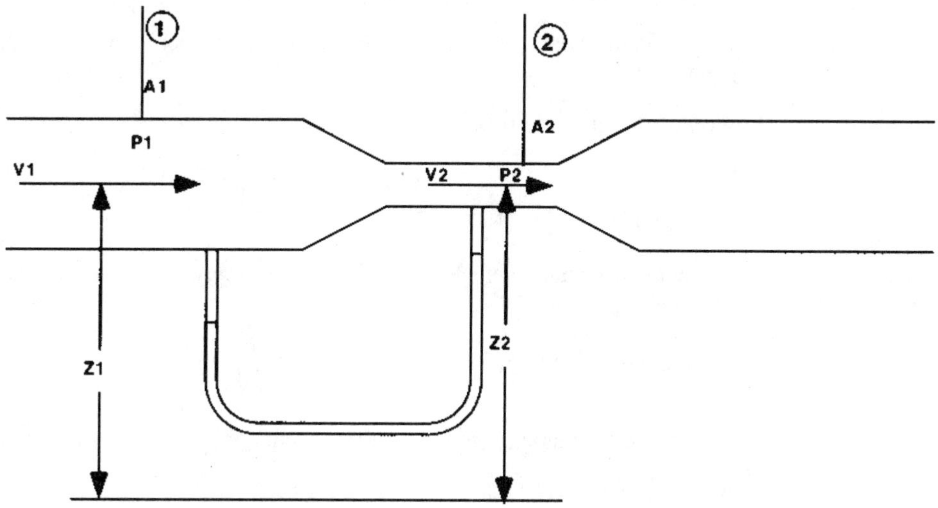

FIGURE 3.2.5 Venturi meter arrangement.

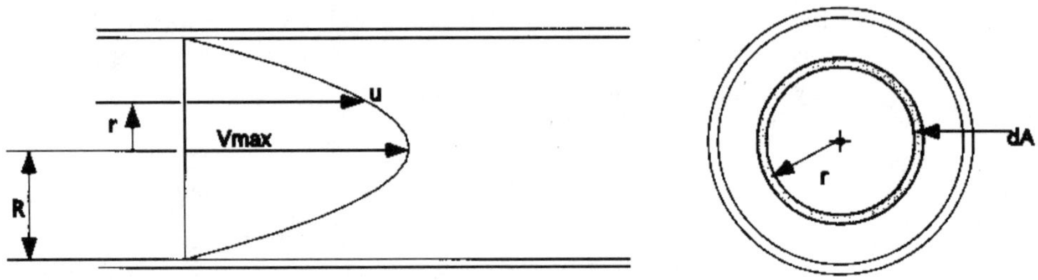

FIGURE 3.2.6 Example velocity distribution — laminar flow

For the elemental area, $dA = 2\pi r dr$:

$$Q = \int_0^R v dA$$

$$\therefore \text{flowrate, } Q = \overline{V}A = \int_0^R V_{\max}\left(1 - \frac{r^2}{R^2}\right) 2\pi r dr$$

$$Q = \overline{V}A = \left(\frac{1}{2}A\right)(V_{\max}) = -\frac{AV_{\max}}{2}$$

One variation of the third method that is often used to compute Q and $\overline{V}$(average velocity) of flow for a gas flowing through a pipe is to measure the velocity at predetermined sections in the flow area. Next, the area of each section is measured. The flow rate is given by:

$$\vec{Q} = \sum \overline{V}_i \times A_i$$

where A_i = Area of section i
V_i = Measured average velocity in section i

The flow in most open channels is turbulent and the velocity profile is typically as shown in Figure 3.2.8.

1. Average top to bottom velocity for a small section given

   a. More precisely by $\overline{V} = \dfrac{V @ \frac{2}{10}D + V @ \frac{8}{10}D}{2}$

   b. An approximate average (top to bottom) velocity (used also for shallow stream flow)

   Average velocity, $\overline{V} \cong V @ \frac{6}{10}D$

2. The maximum velocity (top to bottom) occurs @ $\frac{9}{10}D$ from the bottom of the channel or $\frac{1}{10}D$ from the top.

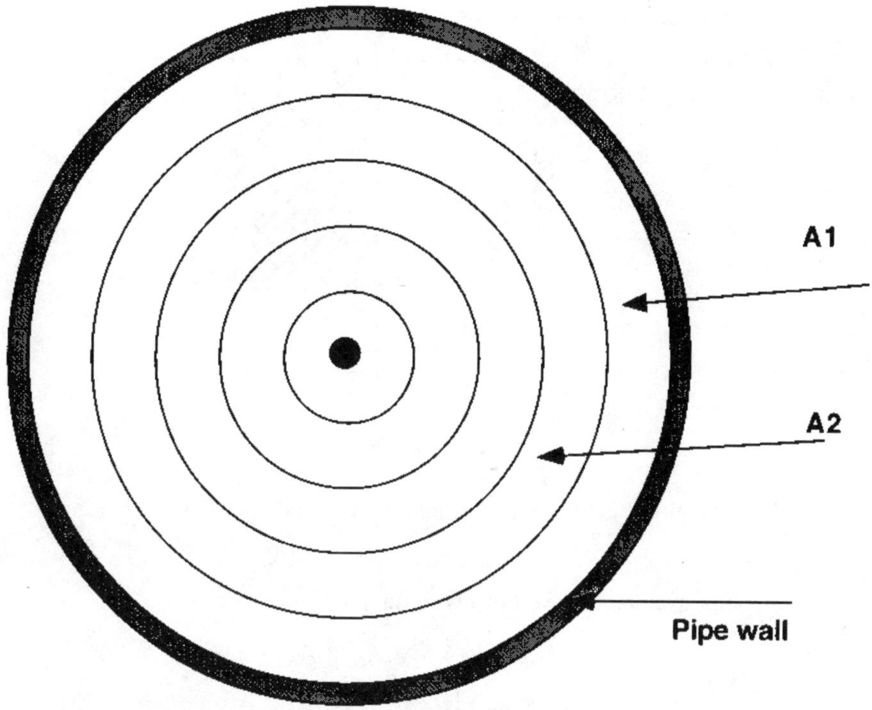

FIGURE 3.2.7 Pipe cross section divided into annular ring segments.

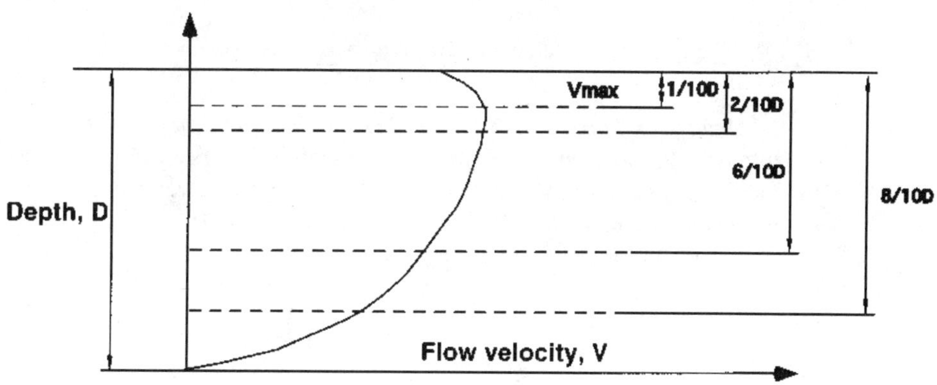

FIGURE 3.2.8 Typical velocity profile in an open channel.

Procedure

1. Depth and velocity measurements are made at each station (A, B, C, D, … G) as shown in Figure 3.2.9.
2. The stations subsequently are used as locations when the average velocity is being determined as midpoints of the sections. See the section marked *B*.

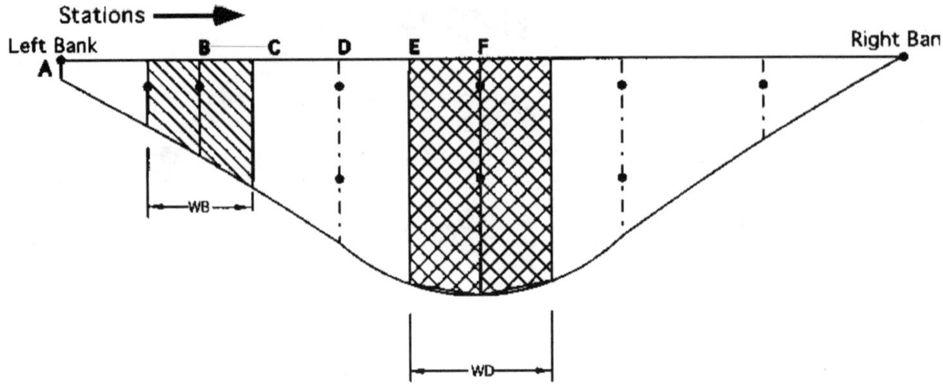

FIGURE 3.2.9 Cross section of an open channel.

3. The elemental flow rate through each section is found:

$$Q_i = A_i \overline{V}_i$$

For example, for sections with midpoints at

$$\text{B: } \overline{V_B} = V_B @ \frac{6}{10} D; \qquad A_B = (Depth_B) \times W_B$$

$$\text{D: } \overline{V_D} = \frac{V_D @ \frac{2}{10} D + V_D @ \frac{8}{10} D}{2}; \qquad A_D = (Depth_D) \times W_D$$

4. The flow rate is calculated from: $Q = \sum_{i-1}^{n} \overline{V}_i A_i$

Momentum Flux (β) and Kinetic Energy Flux (α) Correction Factors. The momentum flux, β, and kinetic energy flux, α, correction factors are used to adjust the energy and momentum equations to account for the fact that average velocity through a section is used in calculations rather than the actual (non-uniform) velocity distribution. The relationships for calculating α and β are given below.

$$\beta = \frac{1}{A} \int \left(\frac{v}{\overline{V}_i} \right)^2 dA$$

$$\alpha = \frac{1}{A} \int_A \left(\frac{v}{\overline{V}_i} \right)^3 dA$$

where v = Velocity function at the cross-section
V = Average velocity across a section
A = Area of flow.

For the same velocity distribution,

$$\alpha \geq \beta \geq 1.00 \quad \text{always.}$$

In addition, for laminar and turbulent flow: $\alpha_{laminar} > \alpha_{turbulent}$ and $\beta_{laminar} > \beta_{turbulent}$.

A correction factor is necessary to account for the non-uniform velocity distribution across a section.

For example, consider the momentum principle for steady, 1-dimensional flow (see Section 3.5 and 3.6 for more detail):

$$\sum F_S + \sum F_B = \int_{cs} \vec{v}\rho\vec{V} \cdot d\vec{A} \approx (\rho\overline{V}_2^2 A)\beta_1 - (\rho\overline{V}_1^2 A)\beta_2$$

where F_S = Surface force
F_B = Body force
$\underline{v}$ = Velocity function
$\overline{V}$ = Average velocity

where $\beta \geq 1$ always, and β_1 and β_2 are the respective correction factors at section 1 and section 2. Notice that β_1 and β_2 are not necessarily equal.

EXAMPLE:
Given that

$$\beta = \frac{1}{A}\int\left(\frac{v}{\overline{V_i}}\right)^2 dA$$

consider the following velocity distribution:

$$U = U_0\left(1 - \frac{r}{R}\right)^m$$

When the flow is laminar and $m \approx \frac{1}{2}$ is assumed and for slightly viscous turbulent flow (assuming $m \approx \frac{1}{7}$):

Typical α and β values are:
For laminar flow:

$$U = U_0\left(1 - \frac{r^2}{R^2}\right)$$

$$\beta = \frac{4}{3} = 1.333 \qquad \text{and} \qquad \alpha = 2.0$$

For turbulent flow:

$$u \approx U_0\left(1 - \frac{r}{R}\right)^m \frac{1}{9} \leq m \leq \frac{1}{5}$$

$$\beta = \frac{(1+m)^2(2+m)^2}{2(1+2m)(2+2m)}$$

$$\alpha = \frac{(1+m)^3(2+m)^3}{4(1+3m)(2+3m)}$$

The turbulent correction factors have the following range of values:
Turbulent flow:

m	$\frac{1}{5}$	$\frac{1}{6}$	$\frac{1}{7}$	$\frac{1}{8}$	$\frac{1}{9}$
β	1.037	1.027	1.020	1.016	1.013
α	1.106	1.077	1.058	1.046	1.037

Notice that these values are close to unity and therefore in most problems they are neglected. The corrections for laminar flow are usually more important as this example illustrates.

3.3 THE BERNOULLI EQUATION

The Bernoulli equation states that for a steady, inviscid (non-viscous, ideal fluid) incompressible flow:

$$P + \frac{1}{2}\rho V^2 + \gamma z = \text{Remains unchanged along a streamline}$$

or

$$\frac{p}{\gamma} + \frac{V^2}{2g} + z = \text{Constant on a streamline}$$

Therefore, if two points along a streamline, 1 and 2 are considered:

$$\frac{p_1}{\gamma} + \frac{V_1^2}{2g} + z_1 = \frac{p_2}{\gamma} + \frac{V_2^2}{2g} + z_2$$

The assumptions are:

1. Virtually viscousless flow—inviscid flow (no friction losses)
2. The flow is steady and thus for every point, the rate of change of velocity with time is zero
3. The specific weight or density of the fluid is not changing—the flow is incompressible
4. The Bernoulli equation is applicable only along a steamline

In order to properly implement the Bernoulli equation it is important to note that

- It is not the energy equation
- It does not consider energy losses due to friction or minor losses due to fittings
- It does not consider changes in the flow area or direction
- It does not apply when machines (pumps and turbines) are between the points on the streamline in which it is applied

3.4 REYNOLDS TRANSPORT THEOREM

This section briefly provides the foundational principles of the Reynolds transport theorem (RTT), also called the *system-to-control volume equation.* The latter name indicates that the RTT provides a means to make a transformation from a *Lagrangian* viewpoint (considering a system of particles) to an *Eulerian* viewpoint (focusing on a portion of space—the control volume). The RTT is important because it provides a vehicle for the derivation of fluid flow equations founded on the laws of conservation of mass, of energy (the first law of thermodynamics), and of linear momentum (Newton's second law). From these, we derive the *continuity equation*, the *energy equation*, and the *momentum equation*.

Definitions:

- *System*: A collection of particles or a given quantity of matter.
- *Extensive property*: A property related to the total mass of a fluid; for example, mass, weight, kinetic energy, momentum, and potential energy.
- *Intensive property*: A property independent of the mass of the fluid and therefore unaffected by the quantity of the fluid, for example, temperature, pressure, velocity, etc. In fact, intensive properties are calculated from the corresponding extensive property divided by the mass of the system. Thus, the intensive property associated with $\text{mass} = \dfrac{\text{Mass}}{\text{Mass}} = 1.$

System-to-Control Volume Transformation (Reynolds Transport Theorem)

Consider the rate of change of an extensive property B, $\dot{B}$. For example, if we considered momentum flux, then:

$$\text{Momentum flux} = \text{Rate of change of momentum} = \frac{d(mv)}{dt} = \dot{m}v$$

1. Now, consider an extensive property of fluid B, where B_{cv} is the amount of B in the control volume.

$$B_{cv} = \int \beta \rho d\forall$$

where $\beta = \dfrac{dB}{dm} = \dfrac{B}{m}$

2. Note that the area vector, $\vec{A}$ is normal to control surface (cs) and points outward.
3. The velocity vector, $\vec{V}$ can be constant across the section(s) or can vary across the sections as shown.
4. The solid line is the control volume (cv), and control surface completely surrounds the control volume (cv).

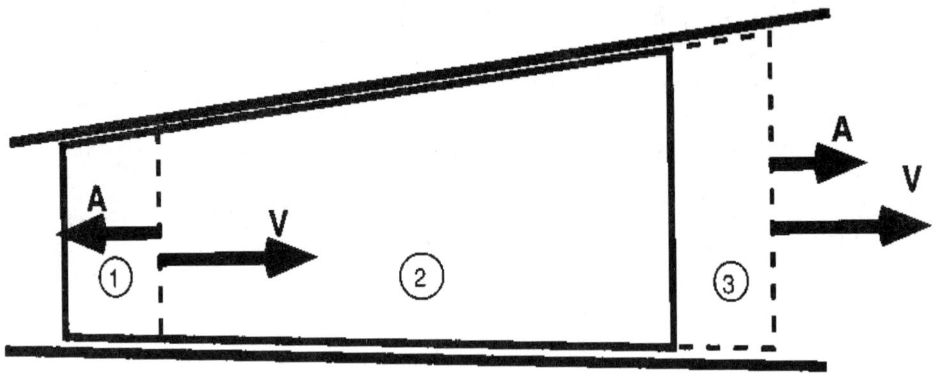

FIGURE 3.4.1 Definition sketch for control volume equation.

5. The rate of flux of B,

$$\mathring{B} = \sum_{cs} \beta\rho\vec{V} \cdot \vec{A} = \int \beta\rho\vec{V} \cdot d\vec{A}$$

$$\frac{dB_{sys}}{dt} = \lim_{\Delta t \to 0}\left[\frac{B_{t+\Delta t} - B_t}{\Delta t}\right] = \text{Rate of change of } B \text{ within the system.}$$

The fluid in the system at time t is equal to fluid in Region [1 + 2] and at time $t + \Delta t$ is equal to the fluid in Region [2 + 3].

That is,

$$\frac{dB_{sys}}{dt} = \lim_{\Delta t \to 0}\left[\frac{(B_2 + B_3)_{t+\Delta t} - (B_1 + B_2)_t}{\Delta t}\right]$$

Rearranged into convenient terms

$$\frac{dB_{sys}}{dt} = \lim_{\Delta t \to 0}\left[\frac{B_{2,t+\Delta t} - B_{2,t}}{\Delta t}\right] + \lim_{\Delta t \to 0}\left[\frac{B_{3,t+\Delta t} - B_{1,t}}{\Delta t}\right]$$

now

$$= \frac{dB_{cv}}{dt} = \frac{d}{dt}\int_{cv} \beta\rho V$$

Therefore, the control volume equation becomes:

1. [for constant velocity across cs]

$$\frac{dB_{sys}}{dt} = \frac{d}{dt}\int_{cv} \beta\rho d\forall + \sum_{cs} \beta\rho\vec{V} \cdot \vec{A} = \mathring{B}$$

The second term on the right side of the above equation is the rate of flux rate of B across the control surface.

The above equation is valid if the velocity vector is constant across every portion of the control surface. Note that only those portions of the control surface through which flow can occur have a non-zero value for flux.

If $\vec{V}$ varies across any portion of the control surface and in particular through the sections through which flow occurs, then

$$\mathring{B}_{cs} = \int_{cv} \beta\rho d\forall + \sum_{cs} \beta\rho\vec{V} \cdot d\vec{A}$$

and the relationship becomes:

2. [for varying V across control surface]

$$\frac{dB_{sys}}{dt} = \frac{d}{dt}\int_{cv} \beta\rho d\forall + \int_{cs} \beta\rho\vec{V} \cdot d\vec{A}$$

Further Explanation of the Control Volume Equation

$$\frac{dB_{sys}}{dt} = \frac{d}{dt}\left(\int_{cv} \beta\rho d\forall\right) + (\beta\rho AV)_{sys} - (\beta\rho AV)_{in}$$

On the right side of the equation:

1. The first term is the rate of change of the extensive property B *within* control volume.
2. The second term is the flux of the extensive property B passing *out* of the control surface.
3. The third term is the flux of the extensive property B passing *into* the control surface.

Finally, the most general equation for the change of B (when $\vec{V}$ is not constant across the control surface) is given as:

$$\frac{dB_{sys}}{dt} = \frac{d}{dt}\left(\iiint_{cv} \beta\rho d\forall\right) + \iint_{cs} \beta\rho V_{\cos\Theta dA_{out}} - \iint_{cs} \beta\rho V_{\cos\Theta dA_{in}}$$

For Steady Flow Conditions

When β and ρ do not vary with time t, then steady flow conditions exist. Thus, the first term on the right side of the above equation is zero.

Additionally, when any of the following special conditions for control volume equation exist:

1. For one-dimensional (constant velocity) steady state:
   In steady state, $\frac{d}{dt}$ and therefore the property B cannot change and so β does not change either. Thus

$$\frac{d}{dt}\int \beta\rho d\forall = 0$$

and therefore

$$\frac{dB_{sys}}{dt} = \sum_{cs} \beta\rho\vec{V} \cdot \vec{A} = (\beta\rho AV)_{out} - (\beta\rho AV)_{in}$$

2. Steady flow conditions with variable $\overset{r}{V}$ across the control surface:

$$\frac{dB_{sys}}{dt} = \int_{cs} \beta\rho\vec{V}_0 \cdot d\vec{A} = \iint_{cs} \beta\rho V\cos\Theta dA_{out} - \iint_{cs} \beta\rho V\cos\Theta dA_{in}$$

3.5 CONSERVATION LAWS IN FLUID FLOW

In this section, a brief presentation is made of each of the conservation laws and the corresponding fluid flow equation is derived. The main idea is to draw attention to the origin and assumptions made in the development of these equations.

3.5.1 The Continuity Principle

The law of conservation of mass states that in a closed system, matter cannot be created nor can it be destroyed. This excludes cases in which nuclear reactions are possible—that is, cases having sources (fission) or sinks (fusion) of mass within the control volume.

For fluid flow, we apply the Reynolds Transport Theorem to a system. Here, the extensive fluid property of interest is the mass, $B = M$

$$\beta = \frac{dB}{dm} = \frac{dm}{dm} = 1$$

therefore

$$\frac{d}{dt}\text{ (Mass)sys} = \frac{d}{dt}\int_{cv} \rho d\forall + \int_{cs} \rho\vec{V} \cdot d\vec{A}$$

But rate of change for mass = 0

$$\therefore \int_{cs} \rho\vec{V} \cdot d\vec{A} = -\frac{d}{dt}\int_{cv} \rho dv \qquad \text{or} \qquad \sum \rho\vec{V} \cdot d\vec{A} = -\frac{d}{dt}\int_{cv} \rho d\forall$$

more generally:

$$\iint_{cs} \rho V\cos\theta dA_{out} - \iint_{cs} \rho V\cos\theta dA_{in} = -\frac{d}{dt}\iiint_{cv} \rho d\forall$$

where θ = Generic angle between the velocity vector and the area vector
cs = Control surface
cv = Control volume

Net rate of outflow of mass = rate of decrease of mass within the control volume:

Steady state: $$\frac{dV}{dt} = 0 \text{ then } \frac{d}{dt}\iiint_{cv} \rho d\forall = 0$$

Constant velocity: $$(\rho AV)_{out} - (\rho AV)_{in} = 0$$

Variable velocity: $$\iint_{cs} \rho V \cos\theta dA_{out} - \iint_{cs} \rho V \cos\theta dA_{in} = 0$$

Continuity Equation for 1-Dimensional Flow

For steady flow: $$\sum_{cs} \rho\vec{V} \cdot \vec{A} = 0$$

$$(\rho AV)_{out} - (\rho AV)_{in} = 0$$

In the figure, we have:

$$\rho_2 A_2 V_2 - \rho_1 A_1 V_1 = 0$$

therefore,

$$\rho_1 V_1 A_1 = \rho_2 V_2 A_2$$

(which makes sense, since we have no sink or source in the system)

If ρ is constant (incompressible flow), then

$$V_1 A_1 = V_2 A_2$$

That is:

$$Q_1 = Q_2$$

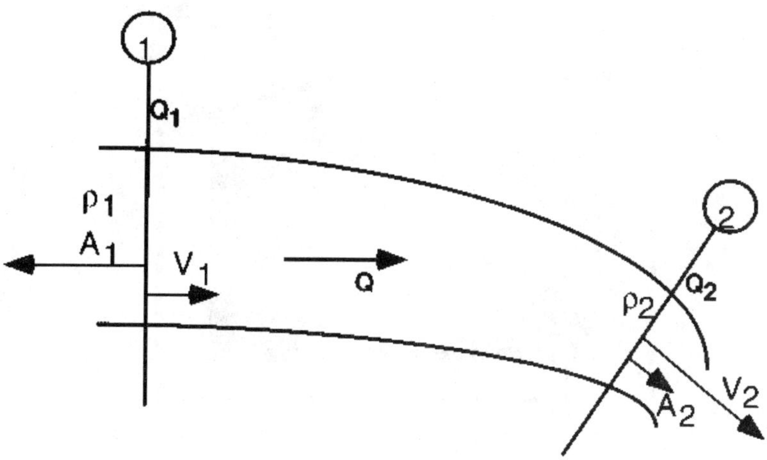

FIGURE 3.5.1.1 Steady 1-dimensional flow through a conduit.

EXAMPLE 3.5.1.1

Kerosene ($\rho = 820$ kg/m^3) is being emptied at a constant rate from a large drum into an open tank. The tank is leaking at a constant rate Q of 400 liters per minute through a large hole at the bottom. If the cross-sectional area A of the jet is 40 cm^2 and the velocity of discharge through the hole (at the bottom) is 6 m/s, determine the rate (liters/minute) at which kerosene is filling up in (or emptying from) the tank.

Solution:

The control surface is drawn to enclose the entire tank. From the Law of Conservation of Mass, the mass flux (rate at which the tank is emptying) is given by:

$$-\frac{d}{dt}\int \rho d\forall = \sum_{\text{cs}} \rho V \cdot A$$

Therefore, the emptying rate (which is the algebraic sum of mass fluxes across the control surface) $= \sum_{\text{cs}} \rho V \cdot A$

$$\text{The net rate of mass flux} = \sum_{\text{cs}} \rho V \cdot A$$

$$= \rho Q_{\text{out}} + \rho V_{\text{in}} \cdot A_{\text{in}}$$

Notice that outflow is positive because the velocity vector and the area vector point in the same direction and inflow is negative because they are opposite in direction.

$$\rho Q_{\text{out}} + \rho V_{\text{in}} \cdot A_{\text{in}} = 820 \times \frac{400 \times 10^{-3}}{60} + 820\left(-6.0 \times \frac{40}{10{,}000}\right)$$

$$= 5.467 - 19.68 \text{kg/s} = -14.213 \text{ kg/s}$$

$$\text{Net rate of loss of kerosene} = -14.213 \text{ kg/s}$$

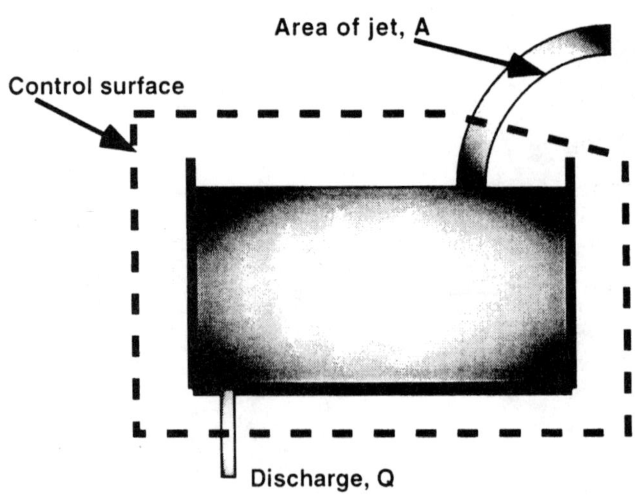

FIGURE E3.5.1.1 Kerosene tank with a leak.

Since the net mass flux (loss) is negative, it means that the tank is actually filling. To obtain the (volumetric) rate of filling, divide the mass flux by the density (ρ) of kerosene.

Therefore the (volumetric) filling rate =

$$q_{\text{fill}} = \frac{\text{mass flux}}{\text{density}} = \frac{14.213 \text{ kg/s}}{820 \text{ kg/m}^3} = 0.01733\text{m}^3 = 0.01733 \times 1000 \times 60 \text{ l/min}$$

$$= 1040 \text{ liters per minute}$$

3.5.2 The Energy Equation

The Energy Equation is developed from the First Law of Thermodynamics (relating heat transfer, work, and energy change in a system.)

The First Law of Thermodynamics states that:

> The rate of heat transfer to a system (Q) minus the rate at which the system does work (W) equals the rate at which the energy of the system is changing $\left(\frac{D}{Dt}\int_{\text{sys}} e\rho d\forall\right)$.

Here

$$e = \text{Specific energy} = \frac{E}{\text{Mass}}$$

where E = Energy in system

Hence:

$$\dot{Q} - \dot{W} = \frac{D}{Dt}\int_{\text{sys}} e\rho d\forall$$

As presented here, this applies to a system (a collection of particles).

Restated, the First Law of Thermodynamics can be written as:

For a given quantity of matter:

$$\Delta E = Q - W \quad \text{(over a given time)}$$

Q = Heat transferred; W = Work done on the surroundings

But

$$\text{Energy } E = E_u + E_k + E_p \equiv \text{Extensive property considered}$$

$$\text{Total Energy} = \text{(Internal)} + \text{(Kinetic)} + \text{(Potential)}$$

In terms of the Reynolds Transport Theorem (RTT), therefore, $B = E$ = Energy.

$$\therefore e = \frac{E_u}{\text{Mass}} + \frac{E_k}{\text{Mass}} + \frac{E_p}{\text{Mass}} = u + e_k + e_p = \text{Specific energy}$$

Recall that the RTT for an extensive property, B and intensive property, is:

$$\frac{DB_{\text{sys}}}{Dt} = \frac{d}{dt}\int_{\text{cv}} \rho d\forall + \int_{\text{cv}} \rho \hat{n} \cdot \vec{V} d_A = \frac{d}{dt}\int_{\text{cv}} \rho d\forall + \int_{\text{cs}} \rho \vec{V} \cdot d\vec{A}$$

where $\hat{n}$ = Unit vector in the direction of the area vector

For the energy in the system:

$$\frac{DE_{\text{sys}}}{dt} = \frac{d}{dt}\int_{\text{cv}} \rho(u + e_k + e_p) d\forall + \int_{\text{cs}} \rho(u + e_k + e_p)\vec{V} \cdot d\vec{A}$$

The energy terms are:

$$e_k = \frac{\frac{1}{2}\Delta m V^2}{\Delta m} = \frac{V^2}{2}; \qquad e_p = \frac{\rho g \Delta\forall_z}{\Delta m} = \frac{\rho g \Delta\forall_z}{\rho\Delta\forall} = g_z$$

⇑(kinetic energy) ⇑(potential energy)

u = Internal energy of fluid particles

Rate of work = $\dot{W}$ (work done on surroundings) and can be divided into two parts:

1. Flow work = Work done by pressure forces as the system moves through space

$$\dot{W}_f = \sum_{\text{cs}} P\vec{V} \cdot \vec{A} = \int_{\text{cs}} P\vec{V} \cdot d\vec{A} = \int_{\text{cs}} \frac{P}{\rho} \rho\vec{V} \cdot d\vec{A}$$

2. Shaft work = Any work that is not flow work
   Includes: Work done through a shaft ($\dot{W}_s$) to:
   a. Put energy into the system (e.g., pump, fan)
   b. Take energy out of system (e.g., turbine)

The Energy Equation becomes:

$$\dot{Q}_s - \dot{W}_s - \int_{\text{cs}} \frac{P}{\rho} \rho\vec{V} \cdot d\vec{A} = \frac{d}{dt}\int_{\text{cs}} \left(\frac{V^2}{V} + gz + u\right)\rho d\forall + \int_{\text{cs}} \left(\frac{V^2}{V} + gz + u\right)\rho\vec{V} \cdot d\vec{A}$$

Combining the third term on the left side and the last term on the right side,

$$\dot{Q}_s - \dot{W}_s = \frac{d}{dt}\int_{\text{cs}} \left(\frac{V^2}{2} + gz + u\right)\rho d\forall + \int_{\text{cs}} \left(\frac{P}{\rho} + \frac{V^2}{V} + gz + u\right)\rho\vec{V} \cdot d\vec{A}$$

When flow is uniform across a cross section, we will use summation symbol (Σ) over the control surface.

For uniform flow:

$$\dot{Q}_s - \dot{W}_s = \frac{d}{dt}\int_{\text{cs}} \left(\frac{V^2}{2} + gz + u\right)\rho d\forall + \sum \rho\left(\frac{P}{\rho} + \frac{V^2}{V} + gz + u\right)\vec{V} \cdot \vec{A}$$

Pipe Flow: Steady, 1-Dimensional, Incompressible Flow. Writing the energy equation between sections [1] and [2], noting that the velocity distributions across sections [1] and [2] are non-uniform:

$$\dot{Q}_s - \dot{W}_s = \int_{cs}\left(\frac{P}{\rho} + \frac{V^2}{2} + gz + u\right)\rho\vec{V}\cdot d\vec{A}$$

specifically for [1] and [2] (Note that $\rho_1 = \rho_2 = \rho$ incompressible flow.)

$$\dot{Q}_s - \dot{W}_s = \int_{A_1}\left(\frac{P_1}{\rho} + gz_1 + u_1\right)\rho V_1 dA_1 + \int_{A_1}\frac{\rho V_1^3}{2}dA_1$$

$$= \int_{A_2}\left(\frac{P_2}{\rho} + gz_2 + u_2\right)\rho V_2 dA_2 + \int_{A_2}\frac{\rho V_2^3}{2}dA_2$$

Now $\dot{W}_2 \equiv$ Rate of shaft work. Shaft work is *usually* either the turbine work ($\dot{W}_t$) or pump work ($\dot{W}_p$).

$$\dot{W}_s = \dot{W}_t - \dot{W}_p$$

takes power from flow ↑ ↑ puts power into flow

Furthermore, since the flow is assumed to be uniform, the streamlines at section [1] and [2] are parallel and straight, hence there is no acceleration normal to the streamlines and so the pressure normal to the streamlines is hydrostatic.

Therefore, the piezometric head

$$\frac{P}{\gamma} + z$$

is constant and so is

$$\frac{P}{\rho} + gz = g\left(\frac{P}{\gamma} + z\right).$$

Additionally the internal energy u is approximately constant across sections [1] and [2].

Hence

$$\frac{P}{\rho} + gz + u \equiv \text{constant}.$$

Therefore,

$$\dot{Q} - \dot{W}_t + \dot{W}_p + \left(\frac{P_1}{\rho} + gz_1 + u_1\right)\int_{A_1}\rho V_1 dA_1 + \int\rho\frac{V_1^3}{2}dA_1$$

$$= \left(\frac{P_2}{\rho} + gz_2 + u_2\right)\int_{A_2}\rho V_2 dA_2 + \int\rho\frac{V_2^3}{2}dA_2$$

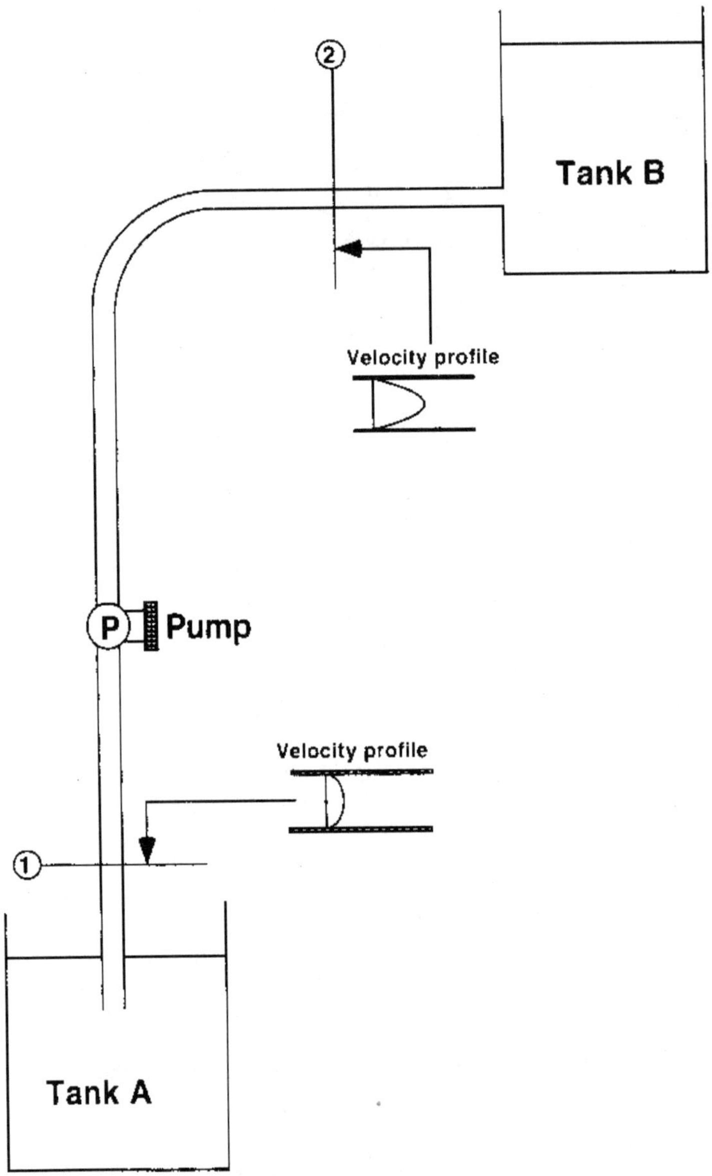

FIGURE 3.5.2.1 Definition sketch for energy equation: steady 1-dimensional uniform flow.

now the term

$$\int \rho \frac{V^3}{2} dA = \alpha \rho \frac{\dot{V}^3}{2} dA = \alpha \rho \frac{\overline{V}^3}{2} A$$

where $\overline{V}$ = Mean velocity across the section
α = Kinetic energy correction factor

Define

$$\dot{m} = \rho \overline{V} A = \int \rho V dA$$

Moving $\dot{Q} - \dot{W}_t$ to the right side of the equation:

$$\dot{W}_p + \left(\frac{P_1}{\rho} + gz_1 + u_1 + \alpha_1 \frac{\overline{V}_1^2}{2}\right)\dot{m} = \dot{W}_t - \dot{Q}_t + \left(\frac{P_2}{\rho} + gz_2 + u_2 + \alpha_2 \frac{\overline{V}_2^2}{2}\right)\dot{m}$$

dividing by $\dot{m}$:

$$\frac{\dot{W}_p}{\dot{m}} + \frac{P_1}{\rho} + gz_1 + u_1 + \alpha_1 \frac{\overline{V}_1^2}{2} = \frac{\dot{W}_t}{\dot{m}} - \frac{\dot{Q}}{\dot{m}} + \frac{P_2}{\rho} + gz_2 + u_2 + \alpha_2 \frac{\overline{V}_2^2}{2}$$

Rearranging the above into *mechanical* and *thermal* parts

$$\frac{\dot{W}_p}{\dot{m}} + \frac{P_1}{\rho} + gz_1 + \alpha_1 \frac{\overline{V}_1^2}{2} = \frac{\dot{W}_t}{\dot{m}} + \frac{P_2}{\rho} + gz_2 + \alpha_2 \frac{\overline{V}_2^2}{2} + \left[u_2 - u_1 - \frac{\dot{Q}}{\dot{m}}\right]$$

mechanical thermal

Finally, if we divide the above by g, we obtain an equation in terms of *length* units:

$$\frac{\dot{W}_p}{\dot{m}g} + \frac{P_1}{\gamma} + z_1 + \alpha_1 \frac{\overline{V}_1^2}{2g} = \frac{\dot{W}_t}{\dot{m}g} + \frac{P_2}{\gamma} + z_2 + \alpha_2 \frac{\overline{V}_2^2}{2g} + \left[\frac{u_2 - u_1}{g} - \frac{\dot{Q}}{\dot{m}g}\right]$$

$\frac{\dot{W}_p}{\dot{m}g} = h_p$, the head supplied by a pump.

The power of the pump, $\dot{W} = \dot{m}gh_p = Q\rho gh_p = Q\gamma h_p$.

Also, $\frac{\dot{W}_t}{\dot{m}g} = h_t$, the head transferred to a turbine.

The power transferred to the turbine, $\dot{W}_t = \dot{m}gh_t = Q\rho gh_t = Q\gamma h_t$.

$$h_p + \frac{P_1}{\gamma} + z_1 + \alpha_1 \frac{\overline{V}_1^2}{2g} = h_t + \frac{P_2}{\gamma} + z_2 + \alpha_2 \frac{\overline{V}_2^2}{2g} + \left[\frac{u_2 - u_1}{g} - \frac{\dot{Q}}{\dot{m}g}\right]$$

↑ h_L, headloss

Headloss is due to heat dissipated from viscous friction. Also, $u_2 - u_1 \cong 0$ and $\overline{V}$ is replaced by V, which is understood to be the average velocity of flow. Therefore, the equation for fluid flow in a pipe for incompressible, steady 1-D flow is:

$$h_p + \frac{P_1}{\gamma} + z_1 + \alpha_1 \frac{V_1^2}{2g} = h_t + \frac{P_2}{\gamma} + z_2 + \alpha_2 \frac{V_2^2}{2g} + h_L$$

Turbines and Pumps. W_p and W_t are measured in horsepower = (550 ft · lb/s) or watts (745.7 watts = 1 horsepower, h_p).

$$1 \text{ watt} = 1\, N \cdot \frac{\text{m}}{\text{s}}$$

3.5.3 Momentum Equation

Let $$B = \text{momentum} = \frac{\text{Momentum}}{\text{Mass}} = \frac{\text{mV}}{\text{m}} = V = \text{Velocity}$$

But: $$\frac{DB_{\text{sys}}}{dt} = \frac{d}{dt}\int_{\text{cs}} \rho d\forall + \int_{\text{cs}} \rho\vec{V} \cdot \vec{A}$$

substituting for B and:

$$\frac{D\text{ (momentum)}}{Dt} = \frac{d}{dt}\int \vec{v}\rho d\forall + \int_{\text{cs}} \vec{v}\rho\left(\vec{V} \cdot \vec{\hat{n}}\right)dA$$

But Newton's second law states that the *time* rate of change of linear momentum = the sum of external forces on a system:

$$\sum\vec{F} = \frac{d}{dt}\int_{\text{cs}} \vec{v}\rho d\forall + \int_{\text{cs}} \vec{v}\rho\left(\vec{V} \cdot \hat{n}\right)dA$$

where $\hat{n}$ = Unit vector in the direction of the area vector,

We use two velocity terms here to distinguish between

$\vec{V}$ ≡ Velocity through control surface with corresponding area vector $\vec{A}$ (a vector)
$\vec{v}$ ≡ Momentum per unit mass (a vector)

Now: $$\sum F = \text{Surface forces} + \text{Body forces}$$
$$= \sum F_S + \sum F_B$$

$\sum F_S$ = Forces transmitted across control surface
$\sum F_B$ = Field force on the mass within the control volume (e.g., gravity, electric or magnetic fields)

$$\therefore \sum\vec{F_S} + \sum\vec{F_B} = \frac{d}{dt}\int_{\text{cs}} \vec{v}\rho d\forall + \sum_{\text{cs}} \vec{v}\rho\left(\vec{V} \cdot A\right)dA$$

If the flow is steady, then there is no change with respect to time: thus

$$\frac{d}{dt}\int_{\text{cs}} \vec{v}\rho dA = 0$$

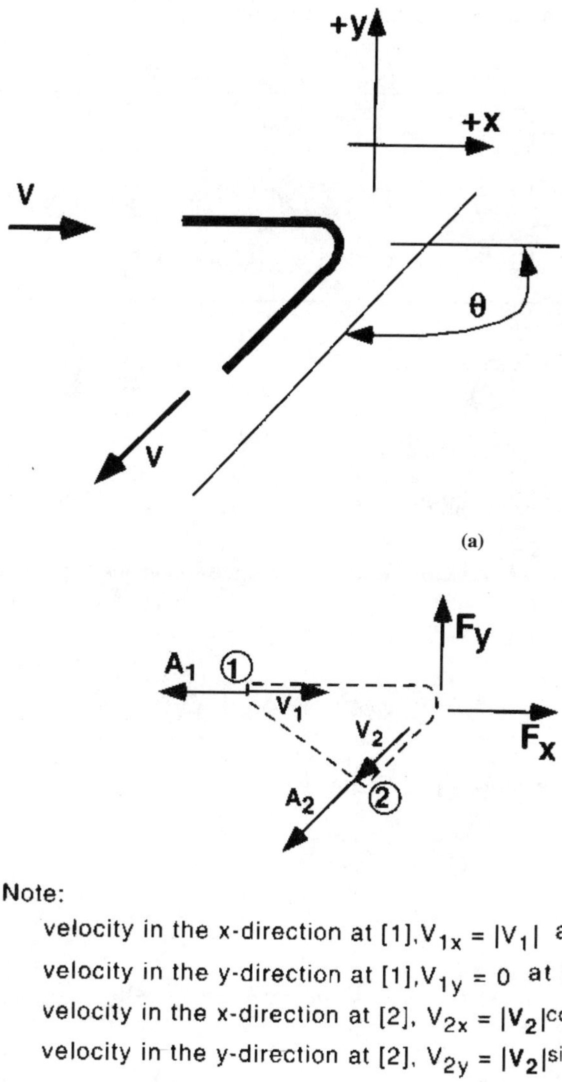

FIGURE 3.5.3.1 Definition diagram for momentum equation.

1. For uniform flow (velocity is uniform across the control surface)

$$\sum \vec{F_S} + \sum \vec{F_B} = \vec{v}\rho\left(\vec{V} \cdot \hat{n}\right)dA$$

2. For any general flow (with non-uniform velocity):

$$\sum \vec{F_S} + \sum \vec{F_B} = \int_{cs} \vec{v}\rho\vec{V} \cdot d\vec{A}$$

For the Cartesian coordinate system, the momentum equations are:

Forces	Uniform flow	General flow	Coordinate
$\sum F_x =$	$\sum_{cs} v_x(\rho\vec{V}\cdot\vec{A}) + \frac{d}{dt}\int_{cs} v_x \rho d\forall$	$\int_{cs} v_x \rho(\vec{V}\cdot\hat{n})dA + \frac{d}{dt}\int_{cv} v_x \rho d\forall$	x
$\sum F_y =$	$\sum_{cs} v_y(\rho\vec{V}\cdot\vec{A}) + \frac{d}{dt}\int_{cv} v_y \rho d\forall$	$\int_{cs} v_y \rho(\vec{V}\cdot\hat{n})dA + \frac{d}{dt}\int_{cv} v_y \rho d\forall$	y
$\sum F_z =$	$\sum_{cs} v_z(\rho\vec{V}\cdot\vec{A}) + \frac{d}{dt}\int_{cv} v_z \rho d\forall$	$\int_{cs} v_z \rho(\vec{V}\cdot\hat{n})dA + \frac{d}{dt}\int_{cv} v_z \rho d\forall$	z

Note that if the flow is steady, then

$$\frac{d}{dt}\int_{cv} v\rho d\forall = 0$$

and the term disappears in all the above equations.

Momentum Equation—Forces on Vanes. The forces on vanes or plates due to a jet will be discussed here.

Considering the x-direction:

$$F_x = \sum_{cs} v_x \rho\vec{V}\cdot\vec{A} + \frac{d}{dt}\int_{cs} \vec{v}_z \rho d\forall$$

For steady flow, the above simplifies to:

$$F_x = \sum_{cs} \vec{v}_x \rho\vec{V}\cdot\vec{A},$$

the force on the jet

$$\sum_{cs} \vec{v}_x \rho\vec{V}\cdot\vec{A} = V_{1x}\rho_1 V_1(-A_1) + V_{2x}\rho_2 V_2 A_2$$

To satisfy continuity:

$$\rho_1 V_1 A_1 = \rho_2 V_2 A_2 = \dot{m}$$

$$\therefore F_x = -V_{1x}(\rho_1 V_1 A_1) + V_{2x}(\rho_2 V_2 A_2) = -V_{1x}\dot{m} + V_{2x}\dot{m} = \dot{m}(V_{2x} - V_{1x})$$

From the above treatment and arrangement, $V_{2x} = V_{1x}\cos\theta$ (for this specific situation). Similarly $F_y = \dot{m}(V_{2y} - V_{1y})$. For this particular case $V_{1y} = 0$.

Generally,

	For steady flow	For this specific case
Forces on the jet	$F_x = \dot{m}(V_{2x} - V_{1x})$	$\dot{m}(V_{1x}\cos\theta - V_{1x})$
	$F_y = \dot{m}(V_{2y} - V_{1y})$	$\dot{m}(V_{2y}\cos\theta - V_{1x})$

$$\text{Finally, the forces on the vane} = -F_x \quad \text{in the } x\text{-direction}$$
$$= -F_y \quad \text{in the } y\text{-direction}$$

If a vane or plate is moving at a velocity V_v then we must use *relative* velocities (relative to the vane or plate). The only condition here is that the plate or vane must not accelerate.

Control volume moves with a velocity V_v

Force on the vane in the x-direction: $$F_{vx} = -\rho\underbrace{(V - V_v)}_{V_x}\left\{\underbrace{(V - V_v)_{2x}}_{V_{2x}} - \left(\underbrace{(V - V_v)_{1x}}_{V_{1x}}\right)\right\}$$

Similarly, forces in the y-direction: $F_{vy} = -\rho(V - V_v)A\{(V - V_v)_{2y} - (V - V_v)_{1y}\}$

Forces on Bends. Assuming uniform and steady flow

$$\left(\frac{d[\,]}{dt} = 0\right)$$

the relevant momentum equation is:

$$\sum F = \sum_{\text{cs}} \vec{v}\rho\vec{V} \cdot \vec{A}$$

In the x-direction:

$$R_x + P_1A_1 - P_2A_2\cos\theta = -V_1A_1(V_{1x}\rho) + V_2A_2(V_{2x}\rho)$$

But $$V_1A_1 = V_2A_2 = Q$$

$$R_x + P_1A_1 - P_2A_2\cos\theta = \rho Q(V_{2x} - V_{1x}) = \rho Q(V_2\cos\theta - V_1) \quad \text{(for this arrangement)}$$

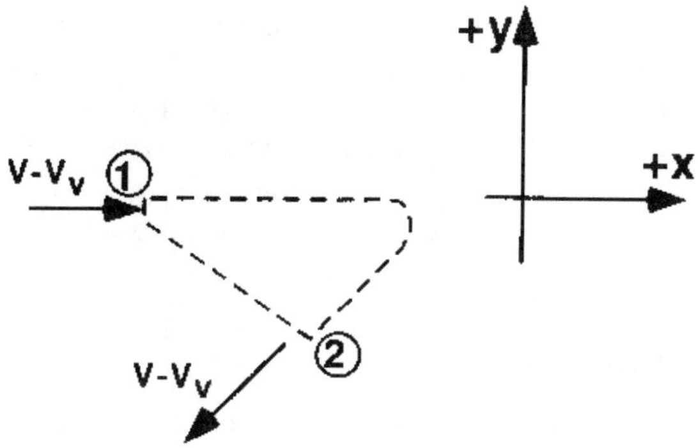

FIGURE 3.5.3.2 A jet impinging on a vane moving at a velocity, V_v.

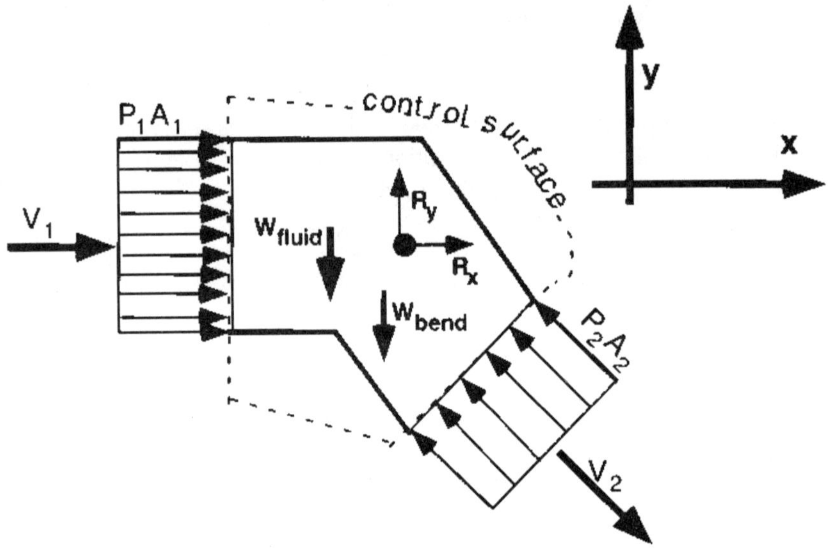

FIGURE 3.5.3.3 Forces on an elbow or bend segment.

In the y-direction:

$$R_y + P_2A_2\sin\theta - W_{bend} - W_{fluid} = \rho Q(V_{2y} - V_{1y}) = \rho Q(V_2\sin\theta) \quad \text{(for this arrangement)}$$

where W_{bend} = Weight of the bend or elbow
W_{fluid} = Weight of the fluid in the bend.

3.6 PRESSURIZED FLOW IN CONDUITS

Introduction

Pressurized flow in conduits refers to fluid flow through conduits of any cross-sectional shape *under pressure*, in contrast to flow in open channels (also called free surface flow where the surface of the liquid is exposed to the surrounding environment at ambient or atmospheric pressure). Generally, it refers to fluid flow in circular conduits. The implementation of the governing equations for fluid flow is presented in this section of the book. The additional discussion augments the basic theory presented in earlier sections. In particular, it is the discussion of the application methods for the equations that were derived (by interpreting the Reynolds Transport Theorem in the context of the conservation laws) in the last section. The emphasis, in this section, is how these equations are used to solve a variety of practical fluid flow problems.

Governing Equations for Fluid Flow

The governing equations for fluid flow are presented and discussed using simple example problems in the following sections. This section provides fundamentalal principles that must be mastered to effectively solve fluid flow problems in the practice of civil engineering.

3.6.1. Continuity Equation (Mass Conservation)

The continuity equation (mass conservation) for two sections along a conduit can be written as follows for the mass flux (mass flowrate):

$$\rho_1 V_1 A_1 = \rho_2 V_2 A_2$$

where ρ_1, ρ_2 = Density of the fluid at sections 1 and 2, respectively
V_1, V_2 = Velocity of sections 1 and 2, respectively
A_1, A_2 = Cross-sectional area at sections 1 and 2, respectively.

The equation states that the mass flux across either section of the conduit is the same. Thus (where $\dot{m}$ is the mass flux):

$$\dot{m}_1 = \dot{m}_2$$

If the flow is assumed to be incompressible, then $\rho_1 = \rho_2$ (which is the case for most normal liquid flow), the equation becomes: $A_1V_1 = A_2V_2$

Flow Relationships in Circular Pipes. Assuming incompressible flow and with reference to Figure 3.6.1.1

Volumetric flowrate, $Q = A_1V_1 = A_2V_2$

$$V_1 = \frac{A_2}{A_1} \qquad V_2 = \frac{Q}{A_1}$$

and

$$V_2 = \frac{A_1}{A_2} \qquad V_1 = \frac{Q}{A_2}$$

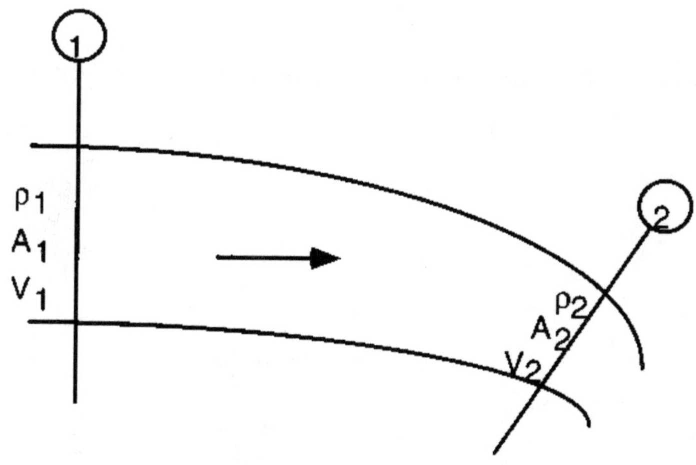

FIGURE 3.6.1.1 Flow through a conduit.

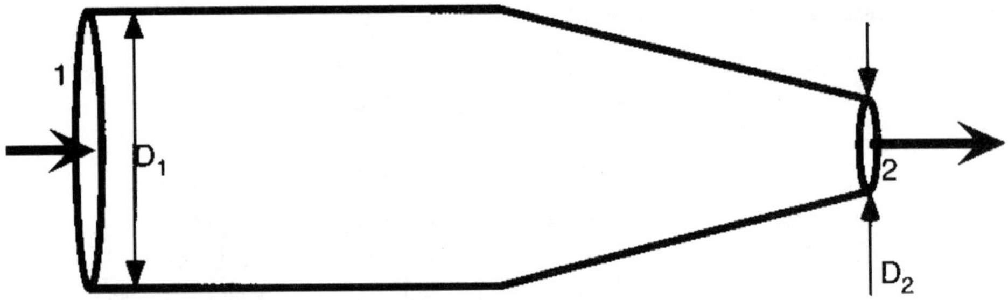

FIGURE 3.6.1.2 Incompressible flow through a tapered circular pipe.

For pipes with a circular cross-section:

$$V_1 = \frac{Q}{A_1} = \frac{4Q}{\pi D_1^2}$$

and

$$V_2 = \frac{Q}{A_2} = \frac{4Q}{\pi D_2^2}$$

Finally,

$$\frac{V_1}{V_2} = \frac{D_2^2}{D_1^2} \qquad \Rightarrow V_1 = \frac{D_2^2}{D_1^2} V_2$$

and

$$\frac{V_2}{V_1} = \frac{D_1^2}{D_2^2} \qquad \Rightarrow V_2 = \frac{D_1^2}{D_2^2} V_1$$

Flowrate Equations for Circular Pipes in Terms of the Flow Velocity, V:

$$Q\left[\frac{\text{m}^3}{\text{s}}\right] = 0.7854 D_m^2 V_m$$

$$Q\left[\frac{\text{ft}^3}{\text{s}}\right] = 0.7854 D^2 V$$

$$Q\left[\frac{\text{ft}^3}{\text{s}}\right] = 0.005454 d^2 V$$

$$q[\text{gpm}] = 352.488 D^2 V$$

$$q[\text{gpm}] = 2.448 d^2 V$$

where d = Diameter in inches
D = Diameter in feet
D_m = Diameter in meters
V = Velocity in ft/s
V_m = Velocity in m/s.

EXAMPLE 3.6.1.1

In the figure below, at A, the diameter of the pipe is 50 cm, the velocity of flow is 10 m/s and the density of the fluid is 4.0 kg/m³. The velocity and diameter at point B are 2 m/s and 150 cm, respectively.

Determine the density of the fluid at point B.

Solution:
By the continuity equation:

$$\rho_A V_A A_A = \rho_B V_B A_B$$

Diameter of pipe at A, $D_A = 0.50$ m; and at B, $D_B = 1.50$ m

$$\rho_A = 4\,\frac{\text{kg}}{\text{m}^3}; \qquad \rho_B = ?$$

$$V_A = 10 \text{ m/s}; \qquad V_B = 2 \text{ m/s}$$

$$\rho_B = \frac{\rho_A V_A A_A}{V_B A_B} = \frac{4 \times 10 \times (\pi \times 0.5^2/4)}{2 \times (\pi \times 1.5^2/4)} = 2.22 \text{ kg\textbackslash m}^3$$

3.6.2 The Energy Equation (Conservation of Energy)

The energy equation that describes the flow between surface of tank #1 and tank #2 in Figure 3.6.2.1 is:

$$Z_1 + \frac{P_1}{\gamma} + \alpha_1 \frac{V_1^2}{2g} + h_{\text{pump}} = Z_2 + \frac{P_2}{\gamma} + \alpha_2 \frac{V_2^2}{2g} + \sum h_f + \sum h_m + h_t$$

where α_1, α_2 = Kinetic energy flux correction factor for sections 1 and 2, respectively
P_1, P_2 = Pressure in flow at sections 1 and 2, respectively
V_1, V_2 = Velocity in flow at sections 1 and 2, respectively
Z_1, Z_2 = Elevation at points 1 and 2, respectively
h_{pump} = Head supplied by the pump

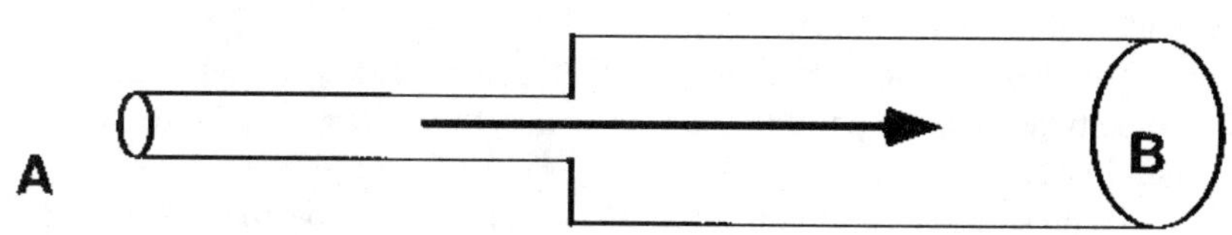

FIGURE E3.6.1.1 Enlarging pipe segment for a compressible flow.

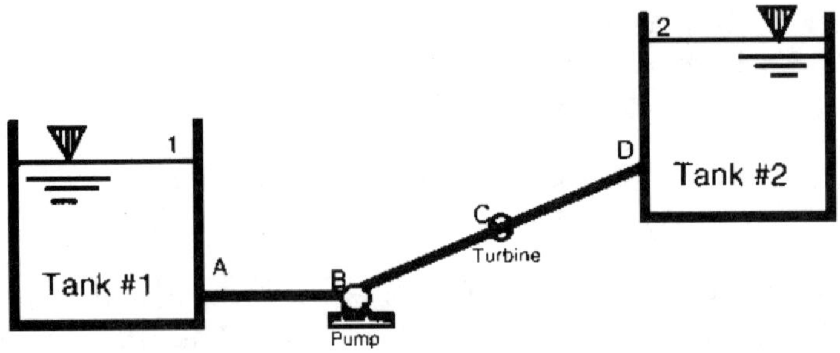

FIGURE 3.6.2.1 Piping system with two reservoirs.

$\sum h_f$ = Total friction headloss between 1 and 2

$\sum h_m$ = Total minor headlosses between 1 and 2

h_t = Energy head converted to mechanical energy by turbine.

If a non-viscous fluid or inviscid fluid flow is assumed, the energy equation simplifies to become the Bernoulli equation. In such a situation, the terms, h_{pump} and $\sum h_f + \sum h_m + h_t$ are dropped because no pumps or turbines can be included in the piping system so as to satisfy the conditions for the Bernoulli equation. Please refer to the stated conditions for the proper application of the Bernoulli equation in Section 3.3.

Hydraulic and Energy Grade Line. With reference to Figure 3.6.2.2 below, the concepts of the Hydraulic Grade Line (HGL) and the Energy Grade Line (EGL) are introduced here. The HGL and EGL are geometric concepts which aid the interpretation of flow concepts in hydraulics. They can be very effective, when properly represented to scale, in the description of the pressure flows and in the location of pumps within piping system. In fact, they can also be help to explain fluid flow concepts in open channel flow.

If a real fluid (one with viscosity) is flowing in a conduit, the EGL must slope downward in the direction of the flow because of the energy loss along the conduit which is due to friction. The level of the EGL at any point in a system, is equal equal to the total energy (head) in the flow. The total energy head is equal to the sum of the pressure head, velocity head (kinetic energy) and the positional (elevation) head. The HGL is the geometric representation of the sum of the pressure head and the elevation head. As can be observed in Figure 3.6.2.2, abrupt changes in the level of the HGL or EGL can occur in a system. Abrupt changes in the height of the EGL or HGL (as can be surmised from Figure 3.6.2.2) occur at locations where a device extracts or adds energy to the flow. In this case, the locations are at the pump and the turbine. Also note the abrupt changes can occur at entrances (from tank to pipe) and exits (from pipe to tank).

The HGL is never higher than the EGL and the difference between the two is equal to the velocity head at the point in the flow. Thus, whenever the velocity head is zero, the HGL and EGL are the same.

To illustrate how the concept of EGL and HGL can be applied, consider the piping system shown in Figure 3.6.2.3. When the HGL and EGL are correctly represented to scale,

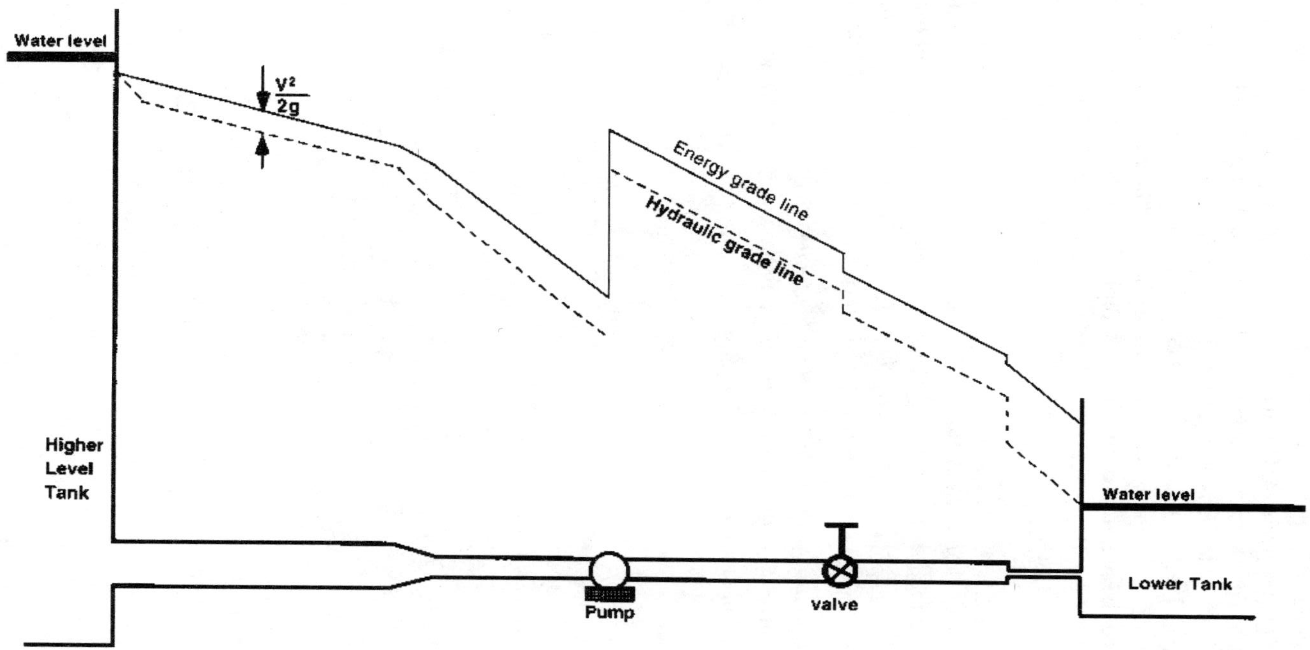

FIGURE 3.6.2.2 Energy and hydraulic grade lines in a hypothetical system.

the distance of the HGL from the pipe (at a given location) provides the measure of the gage pressure in the pipe. If, as in Figure 3.6.2.2, the pipe is *below* the HGL, the gage pressure in the pipe is positive. On the other hand, as in Figure 3.6.2.3, if the pipe is above the HGL, vacuum pressure condition prevails. This can be important in cases where a negative pressure might cause cavitation in the flow. Cavitation can be problematic because it leads to the physical failure of a piping system.

In summary, a properly scaled drawing showing the EGL and HGL can be valuable in determining the pressure conditions in a piping system. Problems with vacuum (negative gage) pressure can be identified easily when the HGL and EGL are provided in the analysis of a piping system.

EXAMPLE 3.6.2.1

Oil (specific weight = 9000 N/m³) is flowing at a rate of 0.40 m³/s in the system shown below. If the total headloss between point [1] and point [2] (sum of friction and minor losses) $h_L = h_f + H_m = 3.2\ V^2/2g$, where V is the velocity in the pipe (diameter = 50 cm), and the pressure at the point [2] is 45 kPa (the elevation is 18 m), determine:

1. *The head, h_p (m) that must be supplied by the pump to the flow*
2. *The power (kW) that must be supplied by the pump to the flow*
3. *The power (kW)Q that must be supplied to the pump (efficiency = 78%).*

Solution:

$$\frac{P_1}{\gamma} + Z_1 + \frac{V_1^2}{2g} + h_p = \frac{P_2}{\gamma} + Z_2 + \frac{V_2^2}{2g} + h_L; \qquad V_1 = 0$$

$$\therefore h_p = \frac{P_2 - P_1}{\gamma} + Z_2 - Z_1 + \frac{V_2^2}{2g} + \frac{3.2V_2^2}{2g}; \qquad P_1 = 0$$

But

$$V_2 = \frac{Q}{A_2} = \frac{0.40}{\left[\dfrac{\pi(0.50)^2}{4}\right]} = 2.037 \text{ m/s}$$

1. Therefore, Pump head

$$h_p = \frac{45 \times 10^3}{9000} + 18 - 10 + \frac{(2.037)^2}{2(9.81)}(1.0 + 3.2) = 13.89 \text{ m}$$

2. Power from pump,

$$W_p = Q\gamma h_p = \left(0.40\ \frac{\text{m}^3}{\text{s}}\right)\left(9000\ \frac{\text{kg}}{\text{m}^3}\right)(13.89 \text{ m}) = 50.00 \text{ kW}$$

3. Power supplied to the pump $= \dfrac{\text{Power supplied by pump to flow}}{\text{Efficiency of pump}}$

$$= \frac{50.00}{0.78} = 64.10\text{kW}$$

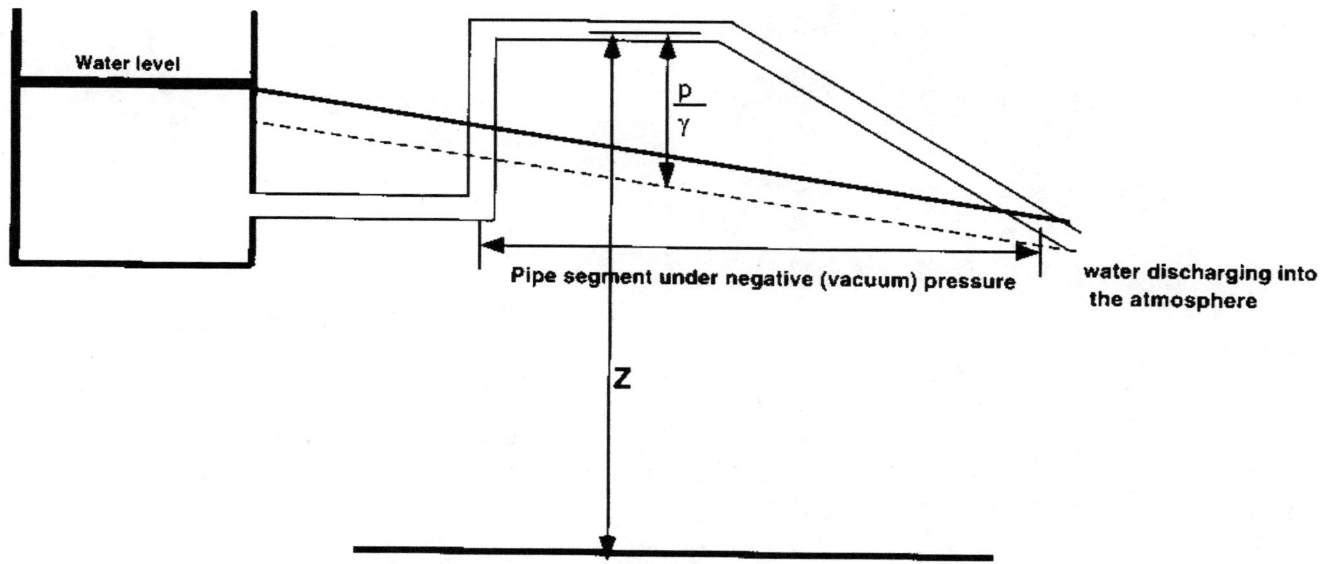

FIGURE 3.6.2.3 Segments of piping system under vacuum pressure.

3.6.3 Momentum Equation

(Derived from Newton's Second Law—Conservation of Linear Momentum)

Referring to Section 3.4, the relationship for the conservation of linear momentum can be stated for a flow to compute the sum of external forces on a system:

$$\sum = \dot{m}\left(\vec{V}_2 - \vec{V}_1\right) = \rho Q\left(\vec{V}_2 - \vec{V}_1\right)$$

where $\sum F$ = The algebraic sum of external forces (surface + body forces)

$\dot{m} = \rho Q$ = Mass flux through the system

$\vec{V}_1, \vec{V}_2$ = Velocity vectors at sections 1 and 2, respectively.

Several cases where the momentum equation have been applied to solve flow problems were presented in Section 3.5. Here, another simple example is used to reinforce the concept.

EXAMPLE 3.6.3.1

A pipe segment (diameter, D = 150 cm) includes a horizontal elbow ($\theta = 60°$) as shown in Figure E3.6.3.1(a). The pipe segment is a part of a piping system used to pump sea water (specific weight = 10,500 N/m^3) at the rate of 7.5 m^3/s to a desalination plant. Assuming the system is such that a pressure of 50 kPa develops in the elbow during pumping and that the elbow weighs 14 kN, determine the x- and y-components of the force (kN) that will be required, in the horizontal plane, to hold the elbow in place. In addition, find the force (kN) that the platform, where the elbow is supported (in the vertical plane), must withstand if the volume of seawater in the elbow is 7.25 m_3.

Solution:

In the 3 orthogonal directions, the sum of the external forces *ExtF* are:

$$\sum ExtF_x = \dot{m}(V_{2x} - V_{1x}) = \rho Q(V_{2x} - V_{1x})$$

$$\sum ExtF_y = \dot{m}(V_{2y} - V_{1y}) = \rho Q(V_{2y} - V_{1y})$$

$$\sum ExtF_z = \dot{m}(V_{2z} - V_{1z}) = \rho Q(V_{2z} - V_{1z})$$

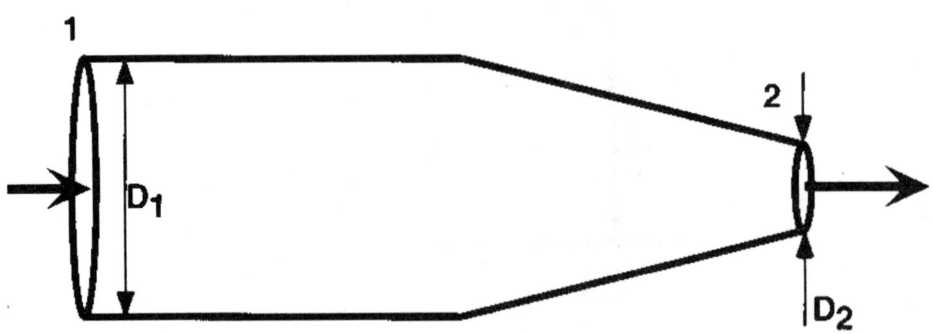

FIGURE 3.6.3.1 Flow through a nozzle.

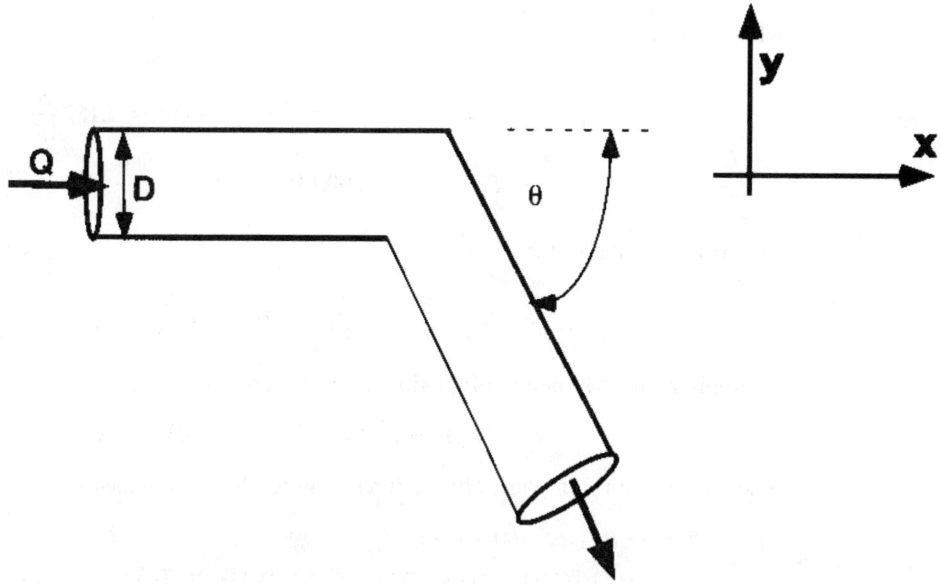

FIGURE E3.6.3.1a Problem definition sketch for example 3.6.3.1.

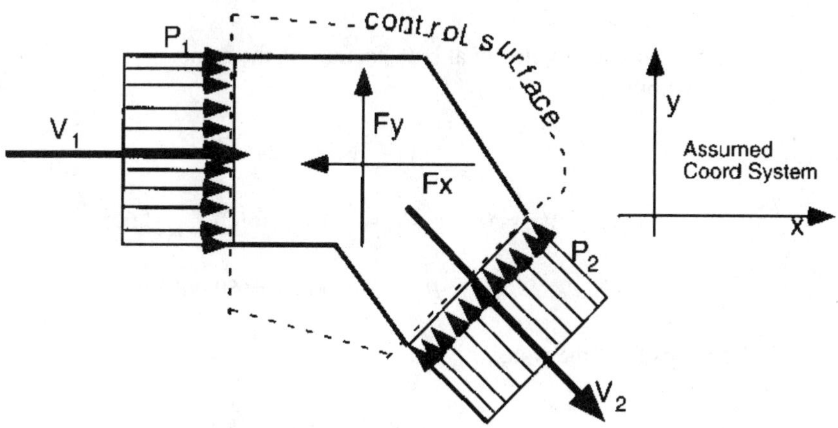

FIGURE E3.6.3.1b Free Body Diagram (FBD) for solving 3.6.3.1.

With reference to the FBD in Figure E3.6.3.1(b), the forces in each orthogonal direction are determined as follows

For the x–direction:

Because the elbow maintains the same diameter as the pipe, the magnitude of velocity at both sections is the same.

By continuity,

$$V_1 = V_2 = \frac{Q}{A_1} = \frac{Q}{A_2} + \frac{7.5\ \frac{\text{m}^3}{\text{s}}}{1.7671\ \text{m}^2} = 4.244\ \frac{\text{m}}{\text{s}}$$

where $A_1, A_2 = \dfrac{\pi(1.5)^2}{4} = 1.7671\ \text{m}^2$

$$V_{1x} = V_1 = 4.244\ \frac{\text{m}}{\text{s}} \qquad V_{2x} = V_2 \cos 60° = 2.122\ \frac{\text{m}}{\text{s}}$$

$$P_1 = P_2 = 50{,}000\ \text{Pa} = 50{,}000\ \frac{\text{N}}{\text{m}^2}$$

The density of the water,

$$\rho = \frac{\gamma}{g} = \frac{10{,}500}{9.81} = 1070.34\ \frac{\text{kg}}{\text{m}^3}$$

Applying the Momentum Equation in the x-direction:

$$-Fx + P_1A_1 - P_2A_2\cos(60°) = \rho Q(V_{2x} - V_{1x})$$

where Fx is the x-component of the force required to restrain the elbow.

$$\begin{aligned} Fx &= P_1A_1 - P_2A_2\cos(60°) - \rho Q(V_{2x} - V_{1x}) \\ &= 50 \times 10^3(1.7671) - 50 \times 10^3(1.7671)(0.50) - 1070.34 \times 7.5 \times (2.122 - 4.244) \\ &\quad + 0.5 \times 50 \times 10^3 \times 1.7671 + 17{,}034.4 = 61210\ \text{N} = 61.21\ \text{kN} \end{aligned}$$

For the y–direction:
The magnitude of velocity at both sections is the same.
By continuity,

$$V_1 = V_2 = 4.244\ \frac{\text{m}}{\text{s}}$$

$$V_{1y} = 0; \qquad V_{2y} = -V_2\sin(60°) = -3.675\ \frac{\text{m}}{\text{s}}$$

$$P_1 = P_2 = 50{,}000\ \text{Pa} = 50{,}000\ \frac{\text{N}}{\text{m}^2}$$

The density of the water,

$$\rho = 1070.34\ \frac{\text{kg}}{\text{m}^3}$$

Applying the Momentum Equation in the y-direction:

$$Fy + P_2A_2\sin(60°) = \rho Q(V_{2y} - V_{1y})$$

where Fy is the y-component of the force required to restrain the elbow.

$$\begin{aligned} Fy &= \rho Q(V_{2y} - V_{1y}) - P_2A_2\sin(60°) \\ &= 1070.34 \times 7.5 \times (-3.675 - 0) - 50 \times 10^3(1.7671)(0.86602) \\ &= -106023\ \text{N} = -106.02\ \text{kN} \end{aligned}$$

Note: Because of the negative sign, Fy is pointing in the opposite direction to the assumed direction (as one would expect).

For the z–direction:

There are no components of velocity or pressure in the z-direction.

Let Fz (pointing vertically upward) be the z-component of the force needed to support the elbow. (Assume that the positive z-direction is vertically upward):

$$\sum ExtFz = \rho Q(V_{2z} - V_{1z}) = 0$$

$$Fz - W_{\text{elbow}} - W_{\text{water}} = 0$$

$$Fz = W_{\text{elbow}} + W_{\text{water}}$$

where W_{elbow} (14 kN) = Weight of the elbow (given)
W_{water} = Weight of the volume of water in the elbow.

But,

$$W_{\text{water}} = \gamma \times \text{Volume} = 10500\,\frac{\text{N}}{\text{m}^3} \times 7.25\ \text{m}^3 = 76.125\ \text{kN}$$

$$\therefore Fz = 14.0\ \text{kN} + 76.125\ \text{kN} = 90.13\ \text{kN}$$

3.6.4 Headloss Due to Pipe Friction

The head loss due to pipe friction can be found by the following methods:

1. Darcy-Weisbach formula:

$$h_f = \frac{fLV^2}{2Dg}$$

where f = Friction factor (from the Moody chart for turbulent flow)
$f = \dfrac{64}{\text{Re}}$ (for laminar flow)

2. By the Chezy-Manning formula,

$$h_L = SL = \frac{V^2 n^2 L}{2.22 R^{4/3}} \quad \text{(for English Units)}$$

$$h_L = SL = \frac{V^2 n^2 L}{R^{4/3}} \quad \text{(for SI Units)}$$

where n = Manning's roughness factor
R = Hydraulic radius = Area of flow\Welted perimeter
V = Velocity
S = Slope of energy grade line (EGL)

3. By the Hazen-Williams formula,

$$h_L = SL = \frac{3.022L}{D^{1.17}}\left(\frac{V}{C}\right)^{1.85} \quad \text{(for English Units)}$$

$$h_L = SL = \frac{6.82L}{D^{1.17}}\left(\frac{BV}{C}\right)^{1.85} \quad \text{(for SI Units)}$$

where C = Hazen-Williams loss coefficient

TABLE 3.6.4.1 Equivalent sand roughness sizes, ϵ

Pipe Material	Equivalent sand roughness sizes, ϵ			
	mm	m	in	ft
Glass, Polyvinyl chloride (PVC) or plastic	$\cong 0$	$\cong 0$	$\cong 0$	$\cong 0$
Drawn tubing, brass, lead copper, bituminous lining	0.0015	0.0000015	0.00006	0.000005
Commercial and welded welded steel	0.045	0.000045	0.00180	0.000150
Riveted steel (few rivets)	0.90	0.000900	0.03600	0.003000
Riveted steel (many rivets)	9.0	0.009000	0.36000	0.030000
Concrete (finished surface)	0.30	0.000300	0.01200	0.001000
Concrete (rough surface)	3.0	0.003000	0.12000	0.010000
Cast iron	0.25	0.000250	0.01000	0.000850
Asphalted (dipped) cast iron	0.12	0.000120	0.00480	0.000400
Galvanized iron	0.15	0.000150	0.00600	0.000500
Wrought iron	0.045	0.000045	0.00180	0.000150
Wood stave (smooth surface)	0.18	0.000180	0.00720	0.000600
Wood stave (rough surface)	0.90	0.000900	0.03600	0.003000

Notes:

- The larger the equivalent sand roughness value, the rougher the pipe and thus the higher the friction head loss.
- In general, the values in the above table are for new, clean pipes. In some cases, therefore, variations (over a range of values) for the pipes must be allowed. As pipes age, the roughness values increase and in particular for small pipes, the internal diameter decreases which results in even higher ϵ/D values, resulting in the larger than proportionate increase in the friction factor, f.

TABLE 3.6.4.2 Manning's Roughness Coefficient, n

Pipe Material	Average or Typical n value
Polyvinyl chloride (PVC) or plastic	0.009
Very smooth pipe (brass or glass)	0.010
Cement-lined ductile iron	0.012
New unlined ductile iron	0.013
Old cast iron	0.020
New cast iron	0.013
Badly corroded cast iron	0.035

TABLE 3.6.4.2 Manning's Roughness Coefficient, *n* (continued)

Pipe Material	Average or Typical *n* value
Wrought iron	0.013
Welded steel	0.014
New riveted steel	0.016
Wood stave	0.016
Vitrified clay pipe (VCP)	0.013
Corrugated metal	0.024
Concrete (rough joints)	0.016
Concrete (dry mix, rough forms)	0.015
Concrete (wet mix, steel forms)	0.012
Concrete (very smooth)	0.011

Note: The larger the *n* value, the rougher the pipe and thus the higher the friction head loss.

TABLE 3.6.4.3 Hazen-Williams Roughness Coefficient, *C*

Pipe Material	Average or Typical C value
Polyvinyl chloride (PVC) or plastic	150
Very smooth pipe (brass or glass, copper or aluminum)	140
Cement-lined ductile iron or cast iron	130
New unlined ductile iron or cast iron	130
Old cast iron	90
Badly corroded cast iron	80
Galvanized iron	120
New riveted steel	110
Welded steel	120
Wood stave	120
Vitrified clay pipe (VCP)	110
Concrete (wood forms)	120
Concrete (steel forms)	140
Brick sewers	100

Notes:

- Hazen-Williams equation applies over a narrow set of conditions for the flow of water in pipes. It is valid for water at a temperature of about 16°C (60° F) and for flow in pipes that are at least 50 mm (2 in.) in diameter and the velocity of flow must be no greater than 3 m/s (10 ft/s).
- The lower the *C* value, the rougher the pipe and thus the higher the friction head loss.
- *C* values also depend on the pipe size (although this is not usually considered in the application of the formula). The larger the pipe size, all things being equal, the greater the *C* value. Average *C* values range from a low of about 80 to a high of about 150.

3.6.5 Minor Losses in Piping Systems

When a flow passes through changes in cross section and/or fittings in the piping system, energy losses occur to these portions of the system. These losses in energy are addressed in two ways. One method is to adjust the length of the piping system by assuming that it is longer. This approach is called the equivalent length method. By this method, a segment of pipe in added to the system to compensate for the headloss introduced. This method, while quite intuitive is not often used.

The second method is to apply a loss coefficient to the system, which is a value (a constant) that is multiplied by the velocity head in the vicinity of the flow. Reference tables for the loss factors are provided in this section. Thus:

$$h_{\text{minor loss}} = K\frac{V^2}{2g}$$

where K (loss coefficient) is used to replace $f\frac{L}{D}$ in the Darcy-Weisbach equation.

Values of K are presented in the tables that follow below.

TABLE 3.6.5.1 Loss Coefficients, K for Pipes Fittings and Valves

Device Class	Type	Condition or State	Loss Coefficient, K
Valves	Gate Valve	fully open	0.20
		$\frac{3}{4}$ open	1.0
		$\frac{1}{2}$ open	5.6
		$\frac{1}{4}$ open	24
	Globe valve	fully open	10
		$\frac{1}{2}$ open	21
	Angle valve	fully open	5.0
	Butterfly valve	fully open	0.4
	Check Valve	fully open (swing type)	2.5
		fully open (lift type)	12.0
		fully open (ball type)	70.0
	Ball valve	fully open	0.10
		$\frac{2}{3}$ open	6.0
		$\frac{1}{3}$ open	21.0
	Foot valve	fully open	15.0
Elbows	90°	regular flanged	0.3
		threaded	1.5

TABLE 3.6.5.1 Loss Coefficients, K for Pipes Fittings and Valves (continued)

Device Class	Type	Condition or State	Loss Coefficient, K
	90° long radius	flanged	0.2
		threaded	0.7
	45° long radius	flanged	0.2
	45° regular	threaded	0.4
Tees	Line flow (straight through)	threaded	0.9
		flanged	0.20
	Branch flow	threaded	2
		flanged	1
180° bend	Return (full U-turn) elbow	threaded	1.5
		flanged	0.20
Union	Joining to segments same D	threaded	0.08

Note: These values vary for the size (diameter) of the pipe. More precise values can be found by considering the specific pipe size. Because of uncertainties regarding specific installation conditions, however, these values are deemed adequate.

TABLE 3.6.5.2 Loss Coefficients, K for Expansions, Contractions, Exit and Entrances

Device Class	Description of Condition	Loss Coefficient, K
Sudden Expansion	A_1 – area of smaller upstream section A_2 – area of larger downstream section	$\left(1.0 - \frac{A_1}{A_2}\right)^2$
Sudden Contraction	A_1 – area of larger upstream section A_2 – area of smaller downstream section	
	$\frac{A_2}{A_1} = 0.0$	0.50
	$\frac{A_2}{A_1} = 0.2$	0.44
	$\frac{A_2}{A_1} = 0.4$	0.30
	$\frac{A_2}{A_1} = 0.6$	0.15
	$\frac{A_2}{A_1} = 0.80$	0.05
	$\frac{A_2}{A_1} = 1.0$	0.00
Entrance: From reservoir to pipe	Re-entrant pipe (protruding into reservoir)	0.80

TABLE 3.6.5.2 Loss Coefficients, *K* for Expansions, Contractions, Exit and Entrances (continued)

Device Class	Description of Condition	Loss Coefficient, *K*
	sharp-edged entrance	0.50
	slightly rounded entrance	0.20
	well-rounded entrance	0.04
Exit:		
From pipe into reservoir	Re-entrant (protruding into reservoir)	1.0
	sharp-edged entrance	1.0
	slightly rounded entrance	1.0
	well-rounded entrance	1.0
90° Mitered bend	without guide vanes	1.1
	with guide vanes	0.20

CHAPTER 4
ENGINEERING HYDROLOGY

This chapter covers engineering hydrology and its applications in the design of engineering structures and management practices. Basic principles in hydrology are addressed. The topics included in the chapter are: the hydrologic cycle, hydrologic budget equation, catchment or watershed scale, watershed characteristics, rainfall distribution, spatial and temporal averaging of rainfall, gage consistency and double-mass analysis, statistics, hydrologic abstractions, the computation of the time of concentration by a variety of methods, unit hydrograph theory and application, the NRCS curve number method, the Rational or Lloyd-Davies formula, and the estimation of evaporation and potential evaporation rates. An example problem in storm water management is used to provide application insights in the latter sections of the chapter.

4.0 DEFINITION AND SCOPE

Hydrology is an earth science that deals with the waters of thc carth. In general, hydrology is subdivided into two major parts—surface water hydrology and groundwater hydrology. This chapter and the sections that follow address surface water hydrology. Some mention of groundwater hydrology is made where it is necessary, but that subject is treated separately in another chapter. Groundwater hydrology deals with waters below the earth's surface; however, there is usually an active interaction between surface waters and groundwater.

Hydrology is one of the earth sciences. The other earth sciences (for the sake of comparison) include climatology (the study of climates), meteorology (a weather science dealing with atmospheric phenomena), geology, geomorphology, sedimentology, geography and oceanography (the study of oceanic life and phenomena).

In other words, hydrology can be viewed as the study of the various forms of water from precipitation to its discharge into the oceans. Thus, it fills the gap between meteorology and oceanography.

To further describe hydrology, we must introduce the concept of the *hydrologic cycle*—a continuous (perpetual) recirculatory path involving the waters of the earth between precipitation (generically called rainfall) and the entry into the oceans.

This process involves water in the atmosphere (usually as water vapor), waters beneath the surface of the earth (groundwater and percolating water, primarily— also referred to as *subsurface water*), and waters on the surface of the earth. The hydrologic cycle is driven by energy from the sun and gravitational forces on the earth. The hydrologic cycle can be said to begin with precipitation in any combination of its many forms (rain, snow, sleet, dew, or hail), and the rainfall can take any of several paths back to the atmosphere as water vapor. The path can include any of the following processes—evaporation, infiltration, percolation, exfiltration, interflow, groundwater recharge, stream flow, and return to the oceans—as it returns to the atmosphere.

Hydrology includes the study of the waters of the earth (often referred to as the *terrestrial phase* of water) by addressing the occurrence, circulation and distribution (spatially and temporally), physical and chemical properties, and interaction and effect of these with man and other living species and the local and global environment. For example, hydrology often deals with runoff, which is the excess water resulting from precipitation and its quantity (volume) flux rate, where it occurs and when, how the water can be used to maintain adequate supplies, and how runoff can be controlled to minimize flooding problems. This book focuses on a particular aspect of hydrology best described as *engineering hydrology*—an applied earth science that uses or applies the principles of hydrology in the solution of engineer-

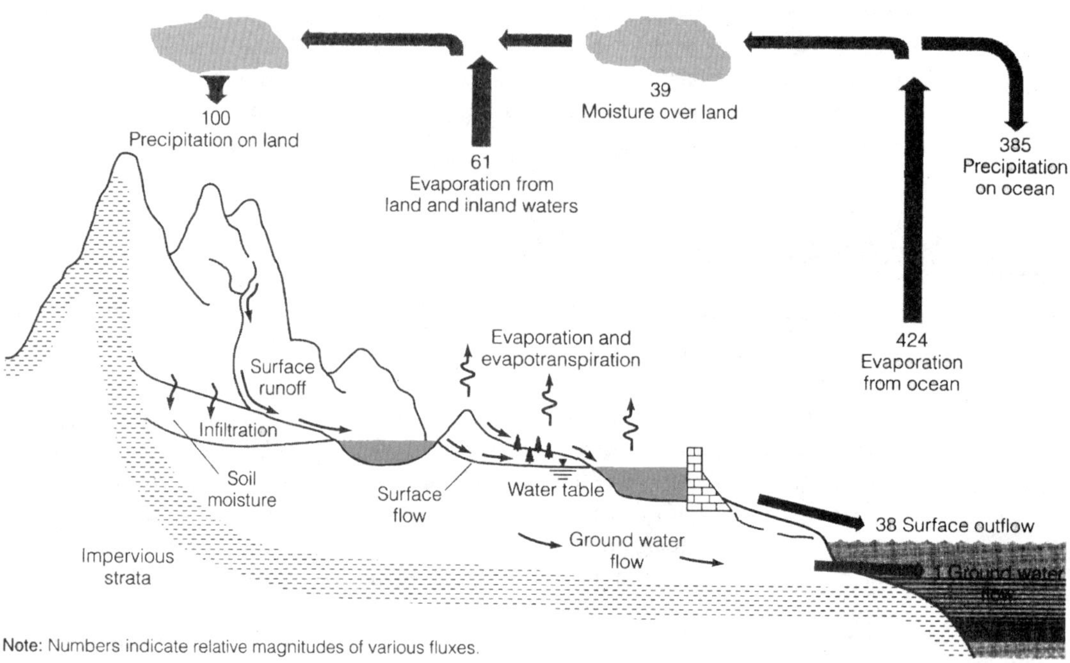

FIGURE 4.0.1 The hydrologic cycle. (*Source:* Applied Hydrology, *Ven Te Chow, David Maidment, Larry W. Mays © 1988, The McGraw-Hill Companies, Reprinted by Permission*)

ing problems that arise from the use of water resources of the earth, and enhancing the quality of life and reducing risks through developing water resources projects.

Engineering hydrology addresses such problems as stormwater management (for example, the design and installation of a storm sewer system to drain the 10-year storm runoff from a proposed highway segment), design of a water supply reservoir, determination of the 50-year storm for a given water resource project, the delineation of the watershed boundary for a given water resources project, flood forecasting, and stream flow forecasting.

4.0.1 Watershed or Catchment

A watershed is an area that will contribute (technically, concentrate) water to a particular common point, called the *basin* or *watershed outlet,* during a rainfall event. The watershed is determined by the shape and topography of the watershed outlet. Generally, water moves downhill, and thus the watershed outlet is at the lowest point in the watershed. The terms *catchment area* or *drainage basin* are used to describe large watersheds. In contrast, the term *catchment*, which describes an area that sheds or concentrates water to an outlet during a rainfall event, is used when relatively small watersheds are considered. The drainage basin typically will encompass several watersheds.

Figure 4.0.2 illustrates the concept of a watershed. To determine the boundary of a watershed, a topographic map that shows contour (ground elevation) lines is necessary. Drainage problems involving a small watershed typically would require 2-foot contours or less. The boundary of the watershed is called the *ridgeline* or the *drainage divide*. Since water runs downhill, runoff or water that is moving over the ground surface will take the path of steepest descent and move in a direction that is perpendicular to the contour lines.

4.0.2 Runoff and Watershed Scale

When a rainfall event occurs, the assumptions that typically are made concerning its spatial and temporal variation can be used to categorize it as small, medium, or large in hydrology. Generally, a rainfall event varies with time and space because its intensity is not constant with time and location. In other words, the rate of rainfall (mm/h) is not constant over the duration of the rainfall event. Furthermore, the rate is not constant from location to location.

The following simplifying assumptions, however, are made for watersheds generically classified as small, midsize, and large.

- Small catchment: Rainfall is assumed to be constant in space and with time.
- Midsize catchment: Rainfall is assumed to be constant in space but varying with time.
- Large catchment: Rainfall is considered to vary in both space and time.

These definitions are summarized in Table 4.0.1 below.

Catchment or watershed scale is taken into account when modeling the rainfall-runoff process.

Small catchments are modeled as in Table 4.1 for which the Rational Method (described later) is applicable.

The rainfall-runoff process of midsize watersheds usually can be modeled using Unit Hydrograph techniques that consider the temporal variation of rainfall.

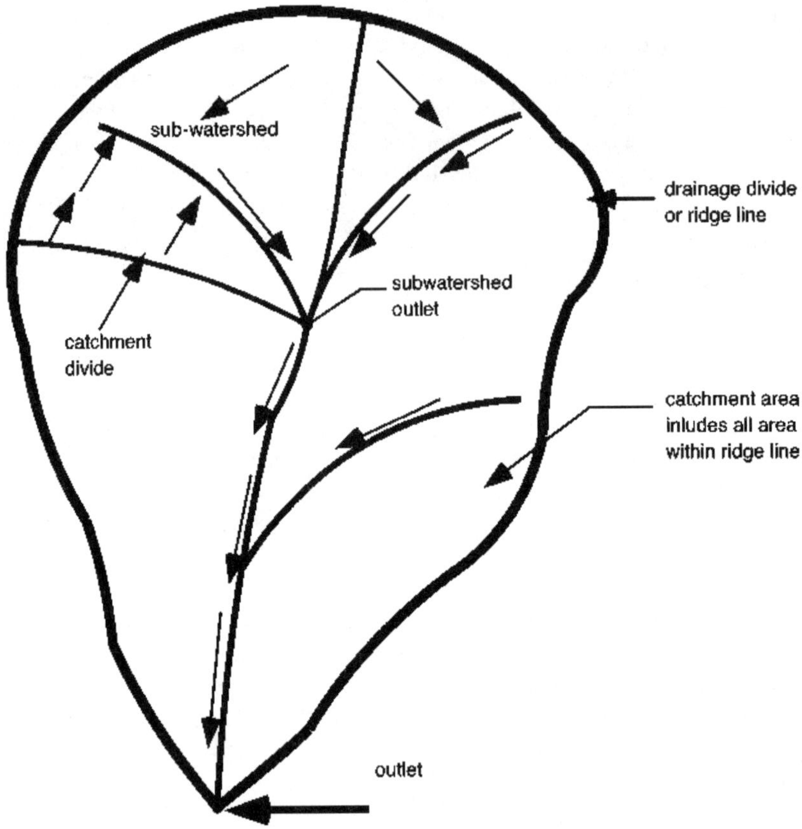

FIGURE 4.0.2 A hypothetical watershed and sub-watershed.

For large watersheds, modeling techniques that take the variation of rainfall in space and time are used. Unit hydrograph techniques assume spatial constancy but can be combined with routing methodologies by breaking up the watershed into convenient sub-units to model large watersheds. Otherwise, modeling techniques that must take into account the temporal and spatial variation of rainfall are used.

TABLE 4.0.1 Watershed Scales with Respect to Rainfall Rate

	Scale	
Small	Midsize	Large
Rainfall rate is constant with time and location	Rainfall rate is constant in space; varies with time	Rainfall rate varies with both space and time.
Area is assumed to be 2.5 km^2 (640 acres) or less, $tc \leq 1-2$ h	area ≤ 25 km^2 , $tc \leq 10$ h	area > 25 km^2 , $tc > 10$ h

4.0.3 List of Formulas

1. Hydrologic budget equation (water balance equation) over a given period for a catchment:

$$\text{Change in storage } \Delta S = P - (I + E + T + Q)$$

or

$$\Delta S = P - (GW + E + T + Q)$$

where ΔS = Change in storage, mm or in.
P = Precipitation, mm or in.
I = Infiltration, mm or in.
E = Evaporation, mm or in.
T = Transportation, mm or in.
Q = Runoff or rainfall excess, mm or in.
GW = Net groundwater outflow, mm or in. (when groundwater is accounted for)

2. Assuming change in storage is zero or known: Runoff (rainfall excess) Q is given by:

$$Q = P - L$$

where P = Precipitation amount, mm or in.
L = Hydrologic abstraction, mm or in.

which includes evaporation, infiltration, transpiration, and any other abstractive processes.

3. Spatial average of rainfall

a. Station average method:
Mean rainfall

$$P_{\text{avg}} = \frac{1}{N}(P_1 + P_2 + \cdots + P_N)$$

where P_i = Rainfall amount of gage station i assuming all N stations are *within* the boundary of the watershed.

b. Isohyetal method:

$$P_{\text{avg}} = \frac{A_1\overline{P}_1 + A_2\overline{P}_2 + \cdots + A_N\overline{P}_N}{A_1 + A_2 + \cdots + A_N}$$

where $\overline{P}_i$ = rainfall amount on isohyet i that represents the average rainfall over area A_i
N = Numbers of sub-areas assumed.

c. Thiessen polygon method:

$$P_{\text{avg}} = \frac{A_1P_1 + A_2P_2 + \cdots + A_NP_N}{A_1 + A_2 + \cdots + A_N}$$

where P_i = Rainfall amount at gage station i for which a Thiessen polygon with area A_i has been determined.
N = Number of Thiessen polygons determined by creating triangles usually equilateral), making perpendicular bisectors, and bisectors, and creating Thiessen polygons.

4. Estimating missing rainfall amount for a gage for a specific rainfall event:
   a. Normal ratio method (for n index stations)

$$P_x = \frac{N_x}{n}\left[\frac{P_1}{N_1} + \frac{P_2}{N_2} + \frac{P_3}{N_3} + \cdots \frac{P_n}{N_n}\right]$$

It is usually sufficient to consider the three closest station*s* to station x (data missing). Assuming stations 1, 2, and 3 are the closest to x:

$$P_x = \frac{N_x}{3}\left[\frac{P_1}{N_1} + \frac{P_2}{N_2} + \frac{P_3}{N_3}\right]$$

Furthermore, if $|N_x - N_i| \leq 0.1N_x$ for all i, then

$$P_x = \frac{P_1 + P_2 + P_3}{3}$$

where P_x = Missing rainfall amount sought for station x
P_i = Precipitation amount for the rainfall event at station i
N_i = Annual rainfall amount for any station, x and i ($\forall i$)
n = Number of index stations (besides station x)

   b. United States National Weather Service method: This is a weighted reciprocal square distance method, assuming an index station in each of the four quadrants (North-East, North-West, South-East and South-West). If more than four index stations are available, only the four stations (one in each of the quadrants) are used.

$$P_x = \frac{W_1P_1 + W_2P_2 + W_3P_3 + W_4P_4}{W_1 + W_2 + W_3 + W_4}$$

where $W_i = \dfrac{1}{L_i^2}$
L_i = Distance of station i from station x.

5. Temporal variation of rainfall:
   Rainfall intensity

$$R = \frac{P}{t_R}$$

where P = Precipitation amount during the period t_R

6. Infiltration indices
   a. ϕ-index

$$\phi\left[\frac{\text{mm}}{\text{hr}}\right] = \frac{P_\phi - Q}{t_\phi},$$

where P_ϕ = Total precipitation amount (mm) when the rainfall intensity is assumed to be greater than the assumed ϕ value
t_ϕ = Total duration in hours (h) when the rainfall intensity is assumed to be greater than the assumed ϕ value
Q = Given amount of the total runoff from the watershed during the rainfall event, mm

b. W-index

$$W\left[\frac{\text{mm}}{\text{hr}}\right] = \frac{P_w - Q - S}{t_w}$$

where P_w = Total precipitation amount, (mm), when the rainfall intensity is assumed to be greater than the assumed W-index value
Q = Runoff amount given, mm
S = Total interception loss plus the depth of surface depression storage, mm
t_w = Total time (duration) corresponding to P_w, hr

7. Instantaneous infiltration equations
   a. Horton's equation

$$f\left[\frac{\text{mm}}{\text{hr}}\right] = fc + (fo - fc)\, e^{-kt}$$

where fo = Initial infiltration rate, mm/hr
fc = Final or terminal/ultimate infiltration
k = Rate of decay of $(fo - fc)$, 1/hr
t = Time elapsed from beginning of rainfall, hr

   b. Philip's equation

$$f\left[\frac{\text{mm}}{\text{hr}}\right] = \frac{1}{2} St^{-\frac{1}{2}} + A$$

where S = Parameter (empirical) related to rate of penetration of the wetting front
A = Parameter related to infiltration that is close to the value of its saturated hydraulic conductivity at the surface of soil
t = Time elapsed, hr

8. Pan evaporation equation

$$E\left[\frac{\text{mm}}{\text{hr}}\right] = K_{pan} E_{pan}$$

where E = Lake evaporation, mm/hr
K_{pan} = Pan coefficient [ ] (a value between 0.0 and 1.0)
E_{pan} = Pan evaporation rate, mm/hr

9. Energy expended for evaporation,

$$Qe\left[\frac{\text{cal}}{\text{cm}^2 - \text{day}} \text{ or } \frac{\text{Ly}}{\text{day}}\right] = \rho LeE$$

where ρ = Density of water, g/cm^3
Le = Latent heat of vaporization, cal/g
E = Evaporation rate, cm/day

10. Penman equation for calculating evaporation rate due to solar radiation flux and mass transfer evaporation

$$E\left[\frac{\text{cm}}{\text{day}}\right] = \frac{\alpha En + Ea}{\alpha + 1}$$

where En = Evaporation rate $\left[\frac{\text{cm}}{\text{day}}\right]$ due to net solar radiation flux Q_n , ly/day

$En = \frac{Qn}{\rho Le}$; ρ and Le are as defined previously
Ea = Mass transfer evaporation rate, cm/day, using Meyer equation or other comparable relationship
α = Temperature-dependent parameter found from reference tables

11. Priestly and Taylor equations for calculating potential evapotranspiration rate for well-watered crops

$$\text{PET}\left[\frac{\text{cm}}{\text{day}}\right] = \frac{1.26\alpha En}{\alpha + 1}$$

where all parameters are as defined previously

12. Time of concentration formulas
    a. Kirpich formula:

$$t_c\ [\text{min}] = \frac{0.0078\ L^{0.77}}{S^{0.385}}$$

where L = Length of channel or representative watershed length, ft
S = Watershed average slope, ft/ft.

In SI units:

$$t_c\ [\text{min}] = \frac{0.3768 L^{0.77}}{S^{0.385}}$$

where L = As defined above, km

    b. Hathaway formula:

$$t_c\ [\text{min}] = \frac{0.8292(L\ n)^{0.467}}{S^{0.234}}$$

where L = Length of main watershed channel or representative watershed length, ft
n = Watershed roughness factor as given in table
S = Watershed average slope

In SI units:

$$T_c\ [\text{min}] = \frac{36.36(L\,n)^{0.487}}{S^{0.234}}$$

where L = As defined above, km

c. Natural Resources Conservation Service (NRCS), formerly Soil Conservation Service (SCS)
From NRCS Lag Equation

$$t_c\ [\text{min}] = \frac{100L^{0.8}[1000 - 9CN]^{0.70}}{1900CN^{0.70}\gamma^{0.50}}$$

where L = Hydraulic length of watershed, usually length to watershed divide, ft
CN = NRCS curve number for the watershed (could be weighted value)
γ = Average watershed slope, %

or in SI units:

$$t_c\ [\text{min}] = \frac{100L^{0.8}(2540 - 22.86\ CN)^{0.70}}{14104\ CN^{0.70}\ S^{0.50}}$$

where L = Hydraulic length, m
S = Average watershed slope, mm.

d. NRCS average velocity equation

$$t_c\ [\text{min}] = \frac{1}{60}\left(\frac{L_1}{V_1} + \frac{L_2}{V_2} + \cdots \frac{L_n}{V_n}\right)$$

where L_i = Length of flow path segment i, (ft or m)
V_i = Average velocity of flow along flow path segment i, (ft/s or m/s)

e. Federal Aviation Administration (for airfields)

$$t_c\ [\text{min}] = \frac{1.8(1.1 - C)L^{0.50}}{Y^{\frac{1}{3}}}$$

where C = Rational Method Runoff Coefficient (could be an average value)
L = Overland flow length, ft
Y = Average ground (surface) slope, %

In SI units:

$$t_c\ [\text{min}] = \frac{3.26(1.1 - C)L^{0.50}}{Y^{\frac{1}{3}}}$$

f. Kinematic wave formula for turbulent flow

$$t_c\ [\text{min}] = \frac{0.94(Ln)^{0.6}}{Ie^{0.4}S^{0.3}}$$

where L = Length of overland flow, ft
n = Manning's surface roughness coefficient for overland flows (ASCE)
I_e = Effective rainfall rate or rainfall excess rate, $\frac{\text{in}}{\text{h}}$
S = Average overland ground slope.

In SI units:

$$t_c\ [\text{min}] = \frac{6.99(Ln)^{0.6}}{Ie^{0.4}S^{0.3}}$$

where L = Length of overland flow, m
n = Manning's n as defined above
I_e = Rainfall excess rate, $\frac{\text{mm}}{\text{h}}$
S = Average overland flow ground slope.

13. Rational formula for calculating runoff from a *small* watershed (area < 640 acres < 2.5 km^2)

$$Q\left[\frac{\text{ft}^3}{\text{s}}\right] = 1.008CIA = CIA$$

(1.008 is usually dropped from the equation)

where C = Runoff coefficient (from reference tables)
I = Rainfall intensity (corresponding to given duration), $\frac{\text{in}}{\text{h}}$
A = Watershed area, acres.

In SI units

$$Q\left[\frac{\text{m}^3}{\text{s}}\right] = 0.278\ CIA$$

where C = As defined above
I = Rainfall intensity, $\frac{\text{mm}}{\text{h}}$
A = Watershed area, km^2.

14. Runoff Curve Number method of the Natural Resources Conservation Service (NRCS), formerly Soil Conservation Service (SCS)
a. Runoff

$$Q\ [\text{in.}] = \frac{(P - 0.2S)^2}{P + 0.80S}$$

where P = Total rainfall amount, in.; $P \geq 0.2$ is assumed
S = Potential maximum retention, in.
$= \frac{1000}{CN} - 10$

CN = Runoff Curve Number (dimensionless) developed by NRCS.

$$1 \leq CN \leq 100$$

The NRCS CN method was developed in English units and corresponding results in SI units are more conveniently determined by performing unit conversions after all calculations have been carried out in English units.

b. An alternative formula for the above is

$$Q\ [\text{in.}] = \frac{[CN(P+2) - 200]^2}{CN\,[CN(P-8) + 800]}$$

for $P \geq \dfrac{200}{CN} - 2$

c. The CN values given in standard tables obtained from NRCS are CN_{II} [for Antecedent Moisture Condition II (AMC II)]. To convert from CN_{II} to CN_I or CN_{III} (AMC I and III, respectively)

$$CN_{\text{I}} = \frac{CN_{\text{II}}}{2.3 - 0.013CN_{\text{II}}}$$

$$CN_{\text{III}} = \frac{CN_{\text{II}}}{0.43 + 0.0057CN_{\text{II}}}$$

CN_{II} values are supplied in the main body of this book and in the appendix.

d. The NRCS unit hydrograph:
Peak flow rate

$$q_P\left[\frac{\text{ft}^3}{\text{s}}\right] = \frac{726\,A\,Q}{t_c}$$

where Q = Effective rainfall amount or runoff, in.
A = Area of the watershed, mi^2
t_c = Time of concentration for the watershed, hrs.

e. Time base and time to peak for NRCS triangular unit hydrograph

$$\text{Time to peak } t_p\ [\text{min}] = \tfrac{2}{3}t_c$$

where $t_c\ [\text{min}] = \dfrac{100L^{0.8}(1000 - 9CN)^{0.70}}{1900CN^{0.70}Y^{0.50}}$

and the time base (triangular hydrograph)

$$t_b\ [\text{min}] = \tfrac{8}{3}t_p = \tfrac{16}{9}t_c$$

15. Probability and statistics

a. Arithmetic mean

$$\overline{X} = \frac{1}{n}\sum_{i=1}^{n} x_i$$

b. Variance and standard deviation
Variance

$$S^2 = \frac{1}{n-1}\left[\sum_{i=1}^{n} x_i^2 - \frac{1}{n}\left(\sum_{i=1}^{n} x_i\right)^2\right]$$

Standard deviation $S = \sqrt{S^2}$

c. Coefficient of variation (measure of relative spread)

$$C_v = \frac{S}{\overline{X}}$$

d. Skew (third moment about the mean)
This is a measure of symmetry about the mean.
Skew

$$g = \frac{n}{(n-1)(n-2)} \sum_{i=1}^{n} (X_i - \overline{X})^3$$

e. Skew coefficient C_s

$$C_s = \frac{g}{S^3} \quad \text{(values can be 0, positive or negative)}$$

g = Skew
S = Standard deviation.

f. Return Period T, Exceedance Probability $P(x \geq X)$, and Risk R

The return period is the mean time that elapses between occurrences of an event of a given size; for example, total rainfall at a location exceeding or matching a certain value.

- $P(x \geq X) = \dfrac{1}{T} = P(X)$

  where T is the return period (usually in years)

- $P(x < X) = 1 - P(x \geq X) = P(\overline{X})$

  $= 1 - \dfrac{1}{T}$ (non-exceedence probability)

- Probability on non-exceedence in n successive periods,

  $$P(\overline{X} \text{ in } n = \left(1 - \frac{1}{T}\right)^n$$

- The risk that an event of a given size (say rainfall exceeding X amount) occurring at least *once* in n periods.

- Risk, $R = 1 - P(x < X)$ in $n = 1 - P(\overline{X})_n$

  $$= \left[1 - \left(1 - \frac{1}{T}\right)^n\right]$$

- Return period, T

$$T = \frac{1}{1 - P(x \leq X)} = \frac{1}{1 - [1 - P(x \geq X)]}$$

g. Standard probability distributions used in hydrology

- Normal distribution

Has two parameters—μ (mean) and σ (standard deviation)
Probability density function (pdf):

$$f(x) = \frac{1}{\sigma\sqrt{2\pi}\, e^{\frac{-(x-4)^2}{2\sigma^2}}}$$

where x is the random variable.

When x is normalized to create the standard normal variate, $Z = \dfrac{x - \mu}{\sigma}$

$$f(Z) = \frac{1}{\sqrt{2\pi}} e^{-\frac{Z^2}{2}}$$

and the cumulative density function (*cdf*):

$$F(Z) = \frac{1}{\sqrt{2\pi}} \int_{-\alpha}^{Z} e^{-\frac{4^2}{2}}$$

- Log Normal Distribution

$$F(x) = \frac{1}{\sigma x\sqrt{2\pi}} e^{-\frac{1}{2}\left(\frac{\ln x - \mu}{\sigma}\right)^2}$$

for > 0

where μ and σ are for $\ln x$

- Log-Person Type III distribution is used to estimate flood flows. The details cannot be provided here because of the summary nature of the formulas provided in this section of the book.

4.1 INTRODUCTION

Engineering hydrology is an applied science that uses or applies hydrologic principles and methods to solve engineering problems that arise from human exploitation of natural resources in the enhancement of the quality of life and to reduce risks from natural or human-induced events. Engineering hydrology provides solutions to such problems as those dealing with storm-water management, water supply reservoir development and construction, flood damage mitigation, water quality improvement, flood forecasting and surface and groundwater interaction. In general, engineering hydrology is an applied science that provides engineering solutions to hydrology and related problems.

Hydrology involves the analysis of historical events (rainfall, temperature, stream flows, rainfall intensity, wind speed and direction, river stages/depth, low flow values, etc.) so inferences can be made about hydrologic processes.

In order to understand natural phenomena and to develop analysis methods to provide engineering solutions to problems, hydrologic data are collected and analyzed. For example, temperature data have been collected and analyzed for several centuries. Thus, predictions about the mean, spread, and range of temperature values for a given location can be provided. The idea here is to use the sample (collected data) to make inferences about the population (the universal of values) for processes of interest to the hydrologist or engineer.

Many hydrologic design methods are based partially (or depend moderately) on theoretical principles; however, the development of the majority of the hydrologic design methods involves a form of data analysis results and procedures. The data analysis method discussed here is probability and statistics.

Hydrology is an inherently empirical science, and as such, to understand hydrologic design methods and procedures properly it is vital to discuss statistical analysis. The discussion of probability and statistics here provides the engineer (or hydrologist) with the background information necessary to understand principles, the foundation for developing new design methods and guidelines, and the limitations and accuracy of the methods developed.

In summary, hydrology is an earth science that deals with the waters of the earth, their occurrence, distribution, and factors that affect or determine such. Design methodologies in hydrology must take into account not only the physical processes that affect the waters of the earth but also the empirical analyses of data describing those processes so that methods and application procedures are developed to match the location and environment in which they must be applied. Often, natural processes are much too complex to describe by basic physical processes or purely analytical methods. Measurements and the analysis of data pertaining to the environment or local are complementary aspects of any good model or procedure for designing or building engineering solutions to hydrologic problems.

In the next section, a brief overview of probability and statistics, in the context of engineering hydrology, is presented.

4.2 BASIC PROBABILITY AND STATISTICS IN HYDROLOGY

The primary objective of statistical (data) analysis is to use the information drawn from a sample (a representative set of data) to make inferences about the population (or the universe of observations). For example, the analysis of 30 years (or periods) of rainfall data could be used to determine the normal or average rainfall in a particular location. In such a case, the implications are that the 30 years of data represent the universe of rainfall data (unobserved past data, observed and recorded data, and all future data) at the locale of interest.

Some basic definitions are presented so the reader is conversant with the terminology and jargon used here and elsewhere.

A *population* is the data universe that includes all data available, unrecorded data in the past, and all future data about a particular process of interest. For example, the rainfall intensity values (in.\hr) for a particular location—a particular watershed, a city, or a county.

A *sample* is a subset (a part) of the population. Usually the sample is chosen or selected so that it is a statistically representative set of the population. In general, the larger the sample, the more representative of the population.

An example of a sample could be the 20 monthly rainfall values measured at the airport location in Greensboro, NC. These values could be used to draw inferences about the monthly values of average rainfall, normal rainfall, and minimum and maximum rainfall amounts. Whether 20 monthly values comprise a representative set is a separate (and important) question that must be addressed by the hydrologist. A sample usually is described as a collection of random data.

Statistical inference is the use of statistical methods (and analysis) so that an investigation that allows inferences or generalizations about a population (using information gained from analyzing representative samples) to be made. Statistical inference is the reason for any interest in hydrological statistical analysis.

A random variable *is a process by which a single number is associated with each outcome of an experiment or a hydrologic process.*

For example, if the process of interest is the accumulated rainfall depth for a two-hour rainfall event in Greensboro, NC, then the associated random variable is "two-hour rainfall depths in Greensboro, NC."

A two-outcome random variable could be whether it rained on any particular day or not. In such a case, the random variable is the value of "rained" or "dry". If X is the random variable, it would have a value for each day (the data was collected) equal to "rained" or "dry."

Probability is a scale used to describe the likelihood of an event that has random occurrences. The scale has a range from 0.0 to 1.0. Zero (0.0) means certain failure (or non-occurrence) and one (1.0) means certain success (assured occurrence).

An event is the occurrence (value of random variable) of a specific value of the random variable. For example, in the two-outcome (rained, dry) experiment described previously, if it rained, the event is a *rainfall event* and the value of the associated random variable is "rained."

Now suppose that during the hurricane season in the coastal portion of North Carolina, the forecast calls for a 90 percent landfall of the eye of a specific hurricane X. This means that, in the past (from the sample of similar hurricanes off the coast of North Carolina), the current conditions show that the several factors that determine the hurricane's path are such that nine out of 10 times the hurricane will make a landfall. This probability implies the concept of *relative frequency*. For example, the probability that a random variable X would take on a particular value K can be represented as:

$$P(x = K)\frac{n}{N}$$

where n = number of outcomes equal to the value K

N = total number observations of the random variable x

Random variables are a major concept in statistics and probability. It is particularly important to understand that probability and statistics are not used to describe a deterministic process. One must assume implicitly that the process being described or studied is a random one. Thus, decisions are made under an uncertain environment or process.

4.2.1 Types of Random Variables

4.2.1.1 Discrete Random Variables. When a random variable can only take one distinct value (usually an integer and usually a finite number), it is called a discrete random variable.

For example, the random variable (rained, dry), which was described previously, has only two outcomes. Another example is the random variable that is associated with the tossing of a die (six outcomes, 1 through 6). For such a random variable, there are six discrete outcomes.

4.2.1.2 Continuous Random Variables. Continuous random variables can take on any value within a continuous range of values, also known as a continuum of values.

Examples:

1. The amount of rainfall that accumulates during a day
2. The 10-day low flow value on a river reach (for every year)
3. The average monthly rainfall in Kernesville, NC.

The distinction between discrete and continuous random variables is important because the method for computing probability values for each is different.

4.2.2 Determination of Probability Values

A probability value is the measure (on the scale 0.0 to 1.0) of the likelihood of a certain outcome of random process. In general, the probability distribution function is used to calculate the probability of a certain outcome of a random process.

A *discrete* random variable uses an associated probability *mass function*, while a continuous random variable uses an associated probability *density function* to describe the likelihood of the outcomes of the random process.

Discrete Random Variable (RV)

4.2.2.1 Probability Distribution Function for Discrete RV. For a discrete random variable, the probability mass function is used to specify the probability that the discrete random variable (say x) equals a specific value (say K_x). Mathematically, this is $P\ (x = K_x)$; sometimes abbreviated as $P(K_x)$.

Every proper probability mass function must have the following properties:

1. $0 \leq P\ (X = K_x) \leq 1.0$ for any value K_x of the random variable X.

   This implies that all probability values must be in the range 0.0 to 1.0.

2. $\sum P\ (X = K_x) = 1$ for all values K_x

This means that the sum of all the probability values for all the possible outcomes of must equal 1. Clearly all possible outcomes of X must have a combined probability of 1.0. For example, the probability of tossing a die and obtaining a value from 1 through 6 is 1.0 (it is assured).

4.2.2.2 The Cumulative Probability Function for Discrete RV. The cumulative probability mass function describes the probability of the discrete random variable over a given range defined as $X \leq K_x$. This is the sum of all probabilities for all outcomes less than or equal to K_x. Therefore for a die (that is not loaded, meaning not compromised in any way) the probability

$$(x \leq 4) = P(x = 1) + P(x = 2) + P(x = 3) + P(x = 4) = \frac{4}{6}$$

Thus $P\,(X = K_x) = \sum P\,(X \leq K_x)$ for all X values not greater than K_x

4.2.3 Continuous Random Variables

4.2.3.1 Probability Density Function. A probability density function *PDF* is used to define the chance or likelihood of the occurrence of a continuous *RV*. In the context of continuous random variables, the probability of the outcome that is a specific value is zero. And thus probability values are non-zero for a range of outcomes. Mathematically,

$$P(x_{K_1} \leq x \leq x_{K_2} = \int_{x_{K_2}}^{x_{K_1}} f(x)\,dx$$

where $f(x)$ is the probability density function.

Note that as the interval $(x_{K_2} - x_{K_1})$ tends to zero (converges to a single value), the probability $P\,(x_{K_1} \leq X \leq X_{K_2}$ becomes zero.

Therefore, the following properties help define the probability density function:

1. $P(x = x_{K_1}) = 0$

   where x_{K_1} is a specific value.

2. $\int_{-\alpha}^{\alpha} f(x)\,dx = 1.0\,;\, f(x) \geq 0$

   The area under the probability density function is equal to 1.00

3. The probability,

$$P(x \leq x_{K_1}) = \int_{-\alpha}^{x_{k_1}} f(x)\,dx$$

$$= 1 - P(x \geq x_{k_1})$$

   Therefore $P\,(x \leq x_{K_1}) + P\,(x \geq x_{K_1}) = 1.0$.

4.2.3.2 Cumulative Density Function. The cumulative density function for a continuous random variable *RV* is the function:

$$F(x_{k_1}) = P(x \leq x_{k_1}) = \int_{-\alpha}^{x_{k_1}} f(\eta)\, d\eta$$

In this case η is the dummy variable of the integration operation.

The cumulative distribution function *F* is non-decreasing and is usually S shaped. For example, the cumulative distribution function for the *normal* distribution is S shaped while the cumulative function for a *uniform* distribution is straight line.

EXAMPLE 4.2.1

Given the following probability density function for the total evaporation E from a lake (in a very humid area) for any day as follows $\left[\frac{in.}{day}\right]$:

$$f(E) \begin{cases} 5 \textit{ for } 0 \leq E \leq 0.20 \left[\frac{\text{in.}}{\text{day}}\right] \\ 0 \text{ otherwise} \end{cases}$$

Determine

1. The probability that the evaporation *E* on a given day $\left[\frac{\text{in.}}{\text{day}}\right]$ will be between 0.1 and 0.15 in.
2. The probability that the evaporation on a particular day will be 0.08 $\frac{\text{in.}}{\text{day}}$
3. The probability that the evaporation will be greater than 0.16 $\frac{\text{in.}}{\text{day}}$
4. The cumulative probability distribution function *F* for the evaporation rates.

Solution

1. $P\,(0.10 \leq E \leq 0.15) = \int_{0.1}^{0.15} f(E)\,dE$

$$= \int_{0.1}^{0.15} 5dE = 5E\Big|_{0.1}^{0.15}$$

$$= 5(0.15 - 0.10) = 0.25$$

Graphically, the probability *density function* is

2. Because *E* is a continuous random variable,

$$P(x = x_{K_1}) = 0$$

Therefore,

$$P(E = 0.08) = 0$$

This is by definition of a probability density function

3. $P(E > 0.16) = \int_{0.16}^{\alpha} f(E)\,dE$

$$= \int_{0.16}^{0.20} f(E)\,dE = \int_{0.16}^{0.20} 5E\,dE$$

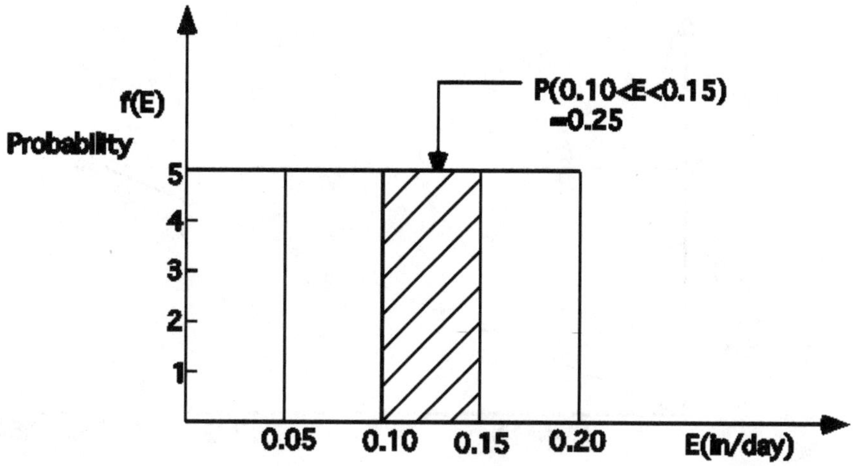

FIGURE E4.2.1a Probability density function.

Since beyond 0.20, $P(x \geq x_{K_1}) = 0$

Therefore,

$$P(E > 0.16) = 5E\Big|_{0.16}^{0.20} = 5(0.20 - 0.16)$$
$$= 0.20$$

Therefore, there is a 20 percent chance that on a particular day the evaporation will exceed 0.16 in.

4. $P(x_{K_1} \leq x \leq x_{K_2}) = F(x_{K_2}) - F(x_{K_1})$

$$F(E) = \int_{-\alpha}^{x_{K_1}} f(E)dE$$
$$= \int_0^{0.20} 5dE\Big|_0^E 5EdE$$
$$F(E) = 5(E - 0) = 5E$$

The CDF for the evaporation process:

$$F(E) \begin{cases} 5E & 0 \leq E \leq 0.20 \\ 0 & \text{otherwise} \end{cases}$$

4.2.4 Plotting Position Formula

A common cumulative probability distribution function used often for extreme events, especially floods, is the plotting position formula. It is used to generate an empirical prob-

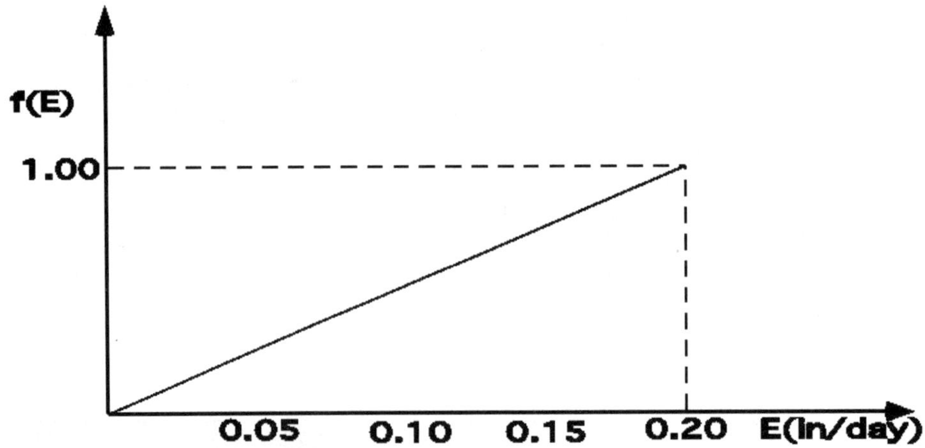

FIGURE E4.2.1b Cumulative probability distribution function.

ability function for random continuous processes. The non-exceedance probability of a random variable x is described as

$$P(x \leq x_k) = \frac{rank}{n + 1}$$

where n observations are ranked in ascending order of magnitude with the smallest observation ranked 1 and the largest ranked n. The above equation gives the non-exceedance probability for X_k when the rank of X_k is known.

In the application of the plotting position formula, the number of observations n must be large and other probability values can be interpolated between the observed values.

4.2.5 Statistical Moments

Statistical moments are measures of central tendency, spread, and skew for a given probability distribution function. The definitions of the common statistical moments are provided here.

Mean (μ). This is the first moment about the origin and it measures the central tendency of a sample. Geometrically, it is the distance of the centroid (of the probability distribution) to the origin. It is also referred to as the *expected value* of the random variable.

Mathematically:

$$\overline{X} = \frac{1}{n} \sum_{i=1}^{n} X_i \quad \text{for a sample}$$

or

$$\mu = \int_{-\alpha}^{+\alpha} f(x)\,x dx \quad \text{for the population if continuous}$$

$$\mu = \sum_{i=1}^{n} x_i\, P(x_i) \quad \text{for the population if } x \text{ is a discrete random variable}$$

Finally, the *median*, another measure of central tendency, is the value that divides a probability distribution into two equal parts. The *mode* is the value of a random variable that has the highest frequency of occurrence.

4.2.6 Variance and Standard Deviation

The variance is the second moment about the mean. The variance indicates the closeness of the values in the sample and hence the population. For a given sample, the variance is given by

$$S^2 = \frac{1}{n+1}\left[\sum_{i=1}^{n} x_i^2 - \frac{1}{n}\left(\sum_{i=1}^{n} x_i^2\right)\right]$$

for the population, the variance is defined for the discrete and continuous random variables as

$$r^2 \text{ (discrete)} = \sum_{i=1}^{n} (x_i - \mu)^2 P(x_i)$$

$$r^2 \text{ (continuous)} = \int_{-\alpha}^{\alpha} (x - \mu)^2 f(x)\, dx$$

for the discrete distribution where each of the n observations have the same chance of occurring.

$$r^2 \text{ (discrete)} = \frac{1}{n}\sum_{i=1}^{w} (x_i - \mu)^2$$

The standard deviation is simply the square root of the variance and gives the spread of the sample or population about the mean, thus

$$S = \sqrt{S^2} \quad \text{or} \quad \sigma = \sqrt{\sigma^2}$$

The *coefficient of variation* measures the relative spread about the mean and is a dimensionless ratio of σ and the mean μ. It is only meaningful for distributions where the mean value is not approximately zero and should be used with care.

Coefficient of variation,

$$C_v = \frac{S}{\bar{x}} \quad \text{or} \quad \frac{\sigma}{\mu}$$

4.2.7 The Skew and Skew Coefficient

The *skew* (γ) is the third moment about the mean and measures the symmetry of the possibility density function (*pdf*). A symmetric distribution (for example, the normal distribution) has a skew of zero. But a non-symmetric distribution can have a positive or negative skew—depending on the orientation of the tail of the *pdf*. If the extreme tail of the *pdf* is to the right of the mean, the skew is positive; otherwise, it is negative

$$\gamma\ (\text{continuous}) = \int_{-\alpha}^{\alpha} (\chi - \mu)^3 f(x)\, dx$$

$$\gamma\ (\text{discrete}) = \sum_{i=1}^{n} (\chi_i - \mu)^3 P(\chi_i)$$

for a population For a given sample, skew,

$$g = \frac{n}{(n-1)(n-2)} \sum_{i=1}^{n} (x_i - \overline{\chi})^3$$

The related skew coefficient C_S is expressed as:

$$C_S = \frac{\gamma}{S^3}, \quad \text{where } S \text{ is the standard deviation.}$$

If $C_S = 0$, the distribution is symmetrical
If $C_S < 0$ the distribution is skewed to the left
If $C_S > 0$ the distribution is skewed to the right.

4.2.8 Standard Probability Distributions

This section presents standard probability distributions used in hydrology. They are:

1. Binomial (discrete distribution)
2. Normal (continuous distribution for *x*)
3. Log normal (continuous for ln *x*)
4. Log Pearson type III distribution.

4.2.8.1 Binomial Distribution or Bernoulli Distribution. Four assumptions are made for this discrete random variable distribution:

1. There are only two possible outcomes.
2. The number of trials *n* is known or set before an experiment is begun.
3. All *n* trials are independent.
4. The probability of either event occurring is the same from trial to trial.

Examples of binomially distributed events:

1. Tossing a coin—you get either a head of a tail. There is no other outcome.
2. Whether a levee will be overtopped or not in any year.

Since there are only two outcomes:
Success (S) or Failure (F)
If prob (success) = Prob (S) = p
Then prob (failure) = $1 - p = q$
Therefore the probability of x successes and exactly $n - x$ failures is:

$$P\ (n \text{ trials}, x \text{ successes}, p) = \binom{n}{x} P^x(1 - P)^{n-x}$$

$$= \frac{n!}{x!(n - x)!} P^x(1 - P)^{n-x}$$

This is the probability density function.
The mean [for n trials, p = prob (success)]

$$\mu = np$$

and the standard deviation,

$$\sigma = \sqrt{np(1 - p)}$$

4.2.8.2 Normal Distribution. The normal distribution has two parameters—μ (mean) and σ (standard deviation). The normal distribution, also called the *Gaussian distribution*, is a continuous distribution. Typically, it is used to describe a continuous random process for which the distribution is bell-shaped in such a way that the mean divides the probability density function into two. If X is a random variable that can be described by the normal distribution, it can be normalized so that the mean of the normalized random variable is 0 and the standard deviation is 1.00. This normalization is done this way:

Say x is normal, then

$$f(x) = \frac{1}{\sigma\sqrt{2\pi}} e - \frac{(x - \mu)^2}{2r^2}$$

then

$$Z = \frac{\chi - \mu}{\sigma}$$

such that

$$\mu_Z = 0 \qquad \text{and} \qquad \sigma_2 = 1.00$$

Thus

$$f(Z) = \frac{1}{\sqrt{2\pi}} e \frac{-Z^2}{2}$$

the cumulative distribution function

$$f(Z) = \frac{1}{\sqrt{2\pi}} \int_{-a}^{z} e \frac{4^2}{2} d\mu$$

where μ is the dummy of the integration process.

$$f(z) = P(\mu \leq Z)$$

EXAMPLE 4.2.2

The mean annual rainfall accumulation at the airport in Greensboro, N.C. is normally distributed and is 42.00 in.\yr with a standard deviation of 5 in.\yr. Determine the probability that drought conditions will occur in a particular year such that total accumulation is only 31 in.\yr.

Solution

Determine the normalized value for the rainfall accumulation

$$Z = \frac{31 - 42}{5} = \frac{-9}{5} = -1.8$$

$$P(Z \leq -1.8) = f(-1.8) = 0.036 \text{ [From Standard Z tables]}$$

The chance of experiencing such a severe drought condition is 3.6 percent in a year.

4.2.8.3 Log-Normal Distribution. The log-normal distribution is a continuous distribution. It has the same characteristics as the normal distribution except that it is the log (usually natural log) of the raw values that is normally distributed.

The effect of taking the natural log of the raw values is that it tends to make a skewed (to the right) set of raw values into a bell-shaped distribution. Hence, the process of taking logs *normalizes* the distribution of log values. In addition, the raw data cannot contain zero or negative values.

The *pdf* is

$$f(x) = \frac{1}{\sigma x \sqrt{2\pi}} e^{-\frac{1}{2}\left(\frac{\ln x - \mu}{\alpha}\right)^2}$$

for

$$x > 0$$

where μ and σ are the mean and standard deviation of $\ln x$

Now suppose that the probability between two values x_1 and x_2 is sought:

$$P(X_1 \leq \chi \leq X_2) = \int^{X_2} \frac{1}{\sigma\sqrt{2\pi}} e^{-\frac{1}{2}\left(\frac{\ln x - \mu}{\sigma}\right)^2} dx$$

But $$F(x_1) = \int_{-\alpha}^{X_1} f(x)dx$$

Therefore $$P(x_1 \le \chi \le x_2) = F\left(\frac{\ln x_2 - \mu}{\sigma}\right) - F\left(\frac{\ln x_1 - \mu}{\sigma}\right)$$

4.2.8.4 Log Pearson III Distribution (LPIII). The LPIII distribution is important in hydrology because it is the approved distribution by the Work Group on Flood Frequency in 1977.

To use the distribution, the log of the flood flow values is used, thus:

Let X = log of observed values.

Then, according to the Work Group on Flood Frequency (Water Resources Council Hydrology Committee), the skew coefficient can be computed by alternate methods.

Please refer to hydrology texts that present a detailed discussion of this topic.

4.2.8.5 Return Period T, Exceedance Probability $P(\chi \le X)$, and Risk R

1. The return period of an event is the mean (average) time that elapses between occurrences of the event.

Therefore $$P(\chi \ge X) = \frac{1}{T}$$

where X = Value to be exceeded
T = Return period of the event (usually in years)

The complementary probability $P(\chi < X)$ is such that $P(x < X) + P(\chi \ge X) = 1.0$

Therefore $P(\chi < X) = 1 - \dfrac{1}{T}$

This is the non-exceedance probability.

The return period, T also can be determined another way.

Since $1 - P(\chi < X)$ is the probability that the event of magnitude X will be equaled or exceeded in any period (year), then

Return period,

$$\begin{aligned} T &= \frac{1}{\text{exceedance probability}} \\ &= \frac{1}{1 - P(\chi < X)} \\ &= \frac{1}{1 - [1 - P(\chi \ge X)} \\ &= \frac{1}{P(\chi \ge X)} \end{aligned}$$

2. If $P(x < X) = 1 - \frac{1}{T}$, the probability that X will not be exceeded in n consecutive periods is

$$P(x < X)^n = \left(1 - \frac{1}{T}\right)^n$$

3. The risk R of an event of given size X occurring at least once—that is, that the event will be equaled or exceeded at least once (for example, that a highway pavement would be flooded at least once in a period of 10 years) is found by:

$$\text{Risk } R = 1 - P(x < X) \text{ over } n \text{ periods}$$
$$= 1 - \left(1 - \frac{1}{T}\right)^n$$

4.3 PRECIPITATION

In general, precipitation is the event whereby water in its many forms (snow, hail, rain, dew, sleet, etc.) falls on the surface of earth. In this section, brief descriptions of the processes that lead to precipitation are presented. The intent is to provide basic background information about the subject.

Precipitation potential exists when moist air is cooled to saturation in the presence of condensation nuclei. The most effective mechanism for cooling a large air mass (rapidly) is called *adiabatic cooling* or *dynamic cooling*. Adiabatic cooling occurs when the pressure of a mass of air is lowered significantly. (A good example of this is the feeling of coolness that results from releasing air at a high pressure into the atmosphere).

In the atmosphere, this happens primarily by *lifting*—the process of moving a mass of air upward in the atmosphere. That is, from a lower level of the atmosphere to a higher level where the pressure is lower, because of the fact that barometric pressure decreases with height (or altitude). Lifting can be achieved in three ways:

1. Horizontal convergence. This occurs when conditions exist in the lower atmosphere such that pressure and wind fields form in such a way as to displace air mass in significant quantities upward. In essence, (moist) air is displaced upward to the upper atmosphere, resulting in cooling.
2. Frontal lifting. This happens when relatively warm air flowing toward a colder denser air mass is wedged upward by the colder air—in this case, a redirection (upward) of the air mass occurs. When the colder (stationary) air displaces the moving warmer air, it is called a *cold front*. If the warmer air is stationary while the colder air mass wedges it upward as the colder air mass moves, it is called a *warm front*.
3. Orographic lifting occurs when moist warm air is forced upward by a mountain or a mountain range, resulting in lifting of the air mass. Orographic lifting is obviously a process that occurs because of relief, and thus this mechanism can only occur in geographic locations with mountainous terrain.

4.3.1 Processes Leading to Precipitation

The occurrence of precipitation is preceded by several processes that must work together to yield appreciable precipitation. Each one of these processes is essential in the production of rainfall (the generic name for all forms of precipitation).

First, the air mass must contain precipitable levels of moisture. Next, the air mass must be cooled to saturation. This saturation point is called the *dew point*—the temperature to which air must be cooled, at constant pressure and water content, to reach saturation. In this case, the important thing is that the air mass must reach the saturation point. When saturation occurs, the water or ice needs surface to condense upon. These are called *condensation nuclei*—particulates of combustion, salt particles from the sea, (ocean) organic salt, and clay minerals (sometimes clouds are seeded with silver iodide to enhance condensation). Condensation nuclei range in diameter from about 0.001 mm to 0.01 mm. If condensation occurs successfully, then droplets form. The droplets must grow into drops of water, balls of ice or snow or to some form of critical mass so that the fall velocity and the size is such that the precipitation can drop through the atmosphere without being flushed upward or disintegrating enough to evaporate. When this happens, precipitation is said to occur. The process of growing, or enlarging the droplets to precipitable size (and mass), is called the *coalescence process*. The mechanism is different for liquid precipitation and solid precipitation.

Sometimes rainfall is referred to as *hydrometeors*, and its various forms include drizzle, rain, snow, hail, and sleet.

4.3.2 Distribution of Rainfall

Rainfall intensity during a rainfall event varies with time. This variation is referred to as the *temporal variation* of rainfall. During a rainfall event, the rate or accumulation over the area of coverage also. This type of rainfall variation is called *spatial variation*. It is important to quantify these distributions of rainfall in time and in space.

A *rainfall hyetograph* is used to quantify the temporal variation of rainfall. And *rainfall isohyets* are used to quantify the spatial variation of rainfall.

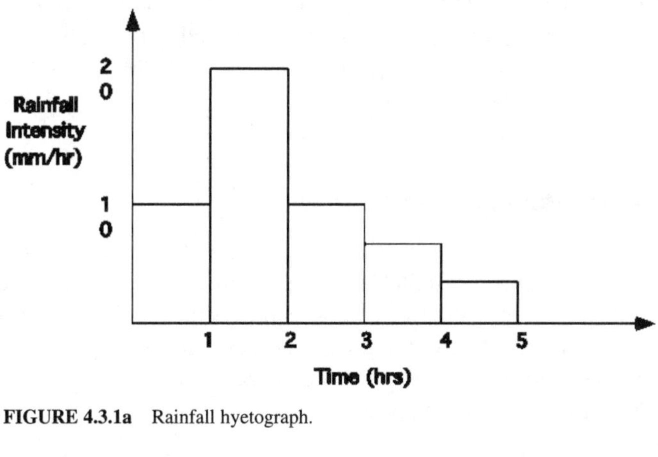

FIGURE 4.3.1a Rainfall hyetograph.

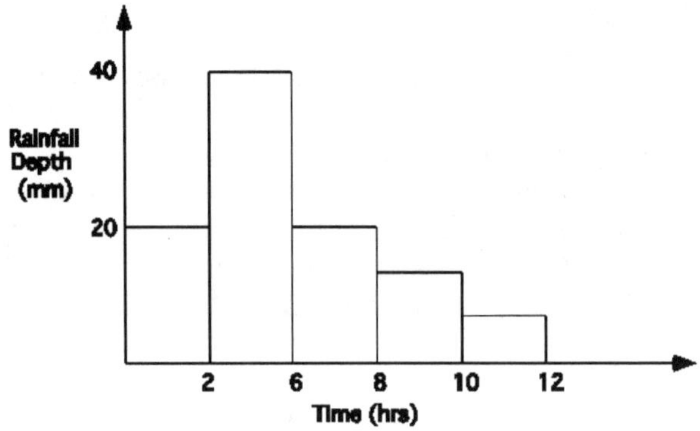

FIGURE 4.3.1b Distribution of rainfall amount during a storm.

Isohyets are lines (similar to contours on a topographic map) that depict the locus of points of equal rainfall depth.

In Figure 4.3.2, the lines show the spatial distribution of rainfall (lines of equal rainfall depth, mm).

In the quantitative description of a rainfall event, a few terms are used.

1. *A rainfall event,* or storm, describes a period of time having measurable and significant rainfall (amount) preceded and followed by periods with no measurable rainfall.
2. The *duration* of a rainfall event is the time from the start to end of the rainfall event or storm. For small watershed, this time period could be in minutes, but for larger catchments it is in hours and could be for several days.
3. The *depth* of a rainfall event is the amount of rainfall considered to be uniformly distributed over the watershed for that particular rainfall event or storm. The depth of rainfall event could be measured in in., cm, or mm.

In general, rainfall depths and duration vary widely. Thus, the temporal and spatial distributions of rainfall depend on the geographic location, the climate, and the season. It can be stated, however, that smaller rainfall depths are more frequently observed than longer rainfall depths. Furthermore, the longer the duration of a rainfall event, the smaller the overall intensity of the storm. Similarly, the larger the watershed (or area over which rainfall is averaged), the smaller the average rainfall.

4.3.3 Calculation of Average Rainfall in a Watershed

Three methods are used to determine the average rainfall in a watershed for a given rainfall event. In most cases, the data collected for determining the average rainfall depth are the recorded rainfall depths measured by rainfall gages distributed over the watershed. This collection of gages often is referred to as the *network of rainfall gages.*

The three methods for averaging the rainfall depth determine how the information collected from the network of rainfall gages is used.

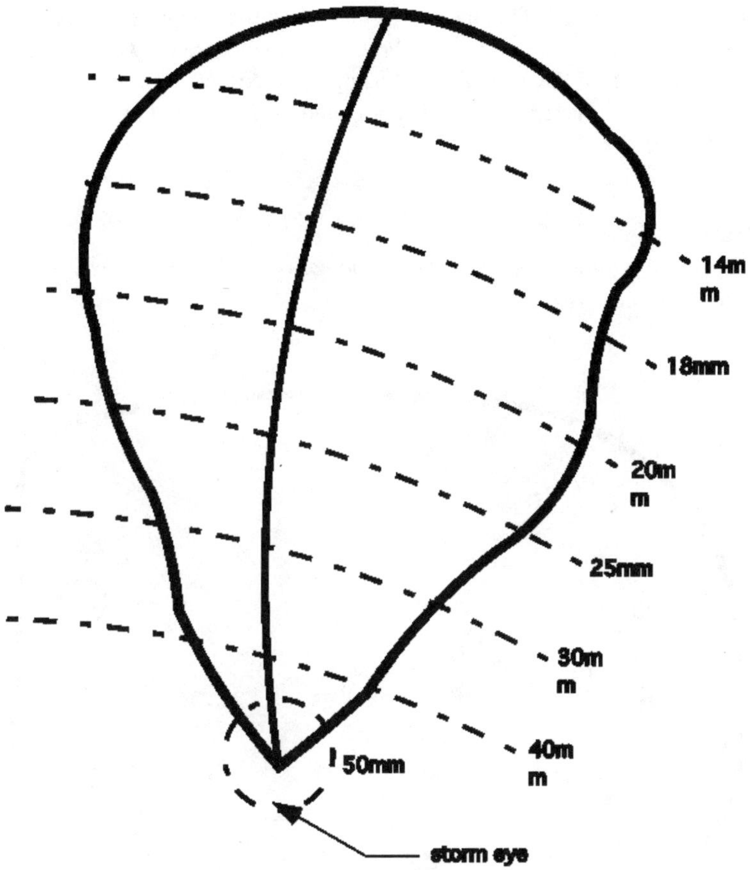

FIGURE 4.3.2 Isohyets showing the spatial distribution of rainfall.

These methods are summarily described below, and an example is used to illustrate the application of each.

1. Station average method. In this method, a simple arithmetic average of the rainfall depths measured by all the rainfall gages that fall *within* the watershed is used to calculate the average spatial rainfall.
2. Thiessen polygon method. Thiessen polygons are drawn by connecting adjacent rainfall gage locations by forming nearly equilateral triangles. The triangles must be such that no internal angle is greater than 90°.

   The sides of the triangles are further bisected at right angles (perpendicular bisectors). These perpendicular bisectors are joined to form polygons.

   Each polygon then will be associated with the rainfall gage that falls within it in such a way that each polygon in the watershed has an associated rainfall station. Finally, weighted average of the rainfall values (with the weight for each station equal to the relative area of its associated polygon) is the average precipitation.

3. The isohyetal method. This method is much like the previous one except that instead of using a mechanistic polygon method, the isohyets are drawn to portray the true spatial variation of the rainfall event using the recorded values at each rainfall station. The spatial average rainfall is computed as a weighted average value as described below.

EXAMPLE 4.3.1

Determine the spatial acreage rainfall for the network of rainfall gages shown in Figure E4.3.1a using the:

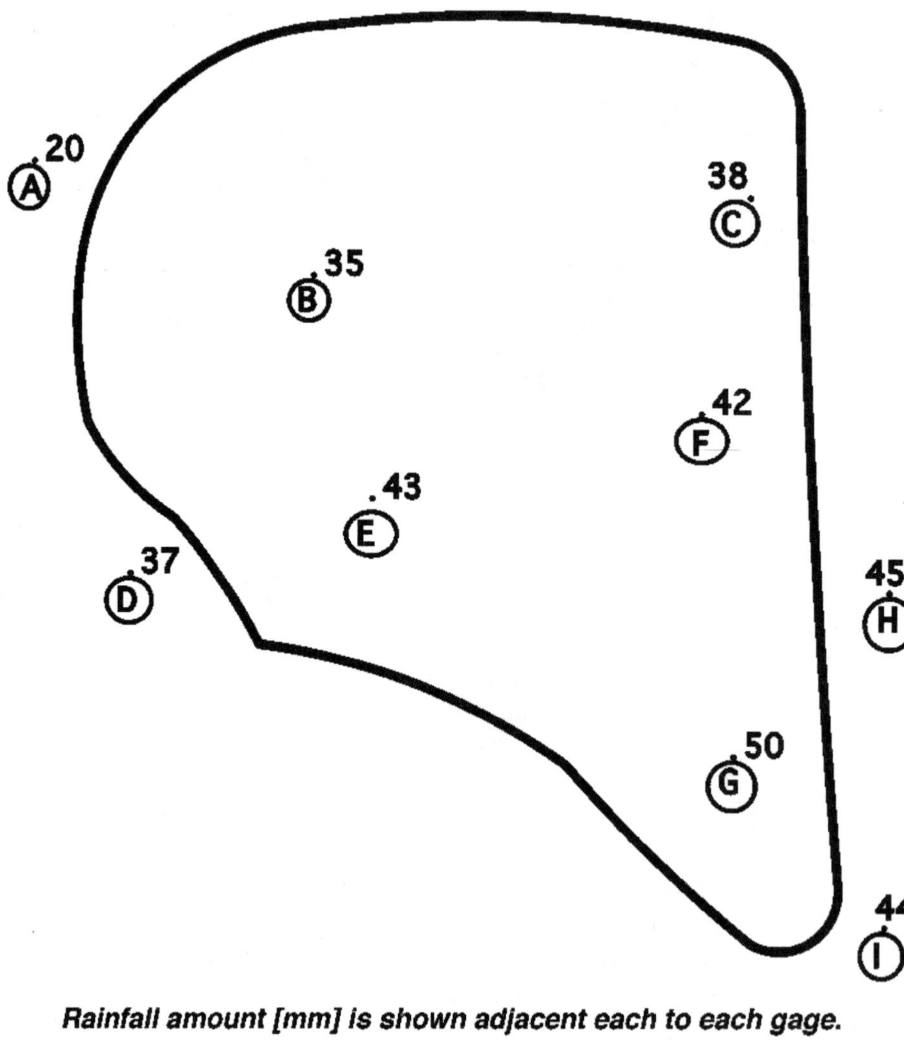

FIGURE E4.3.1a Rainfall stations in the gage network of Example 4.3.1.

1. *Station average method*
2. *Thiessen polygon method*
3. *Isohyetal method*

The values shown on the figure are in mm.

Solution (1)
Using the station acreage method

$$\text{Average rainfall} = \frac{35 + 38 + 43 + 42 + 50}{5} = 41.6 \text{ mm}$$

Observe that checked (✓) stations are the only ones used to determine the average spatial rainfall because they fall within the watershed boundary.

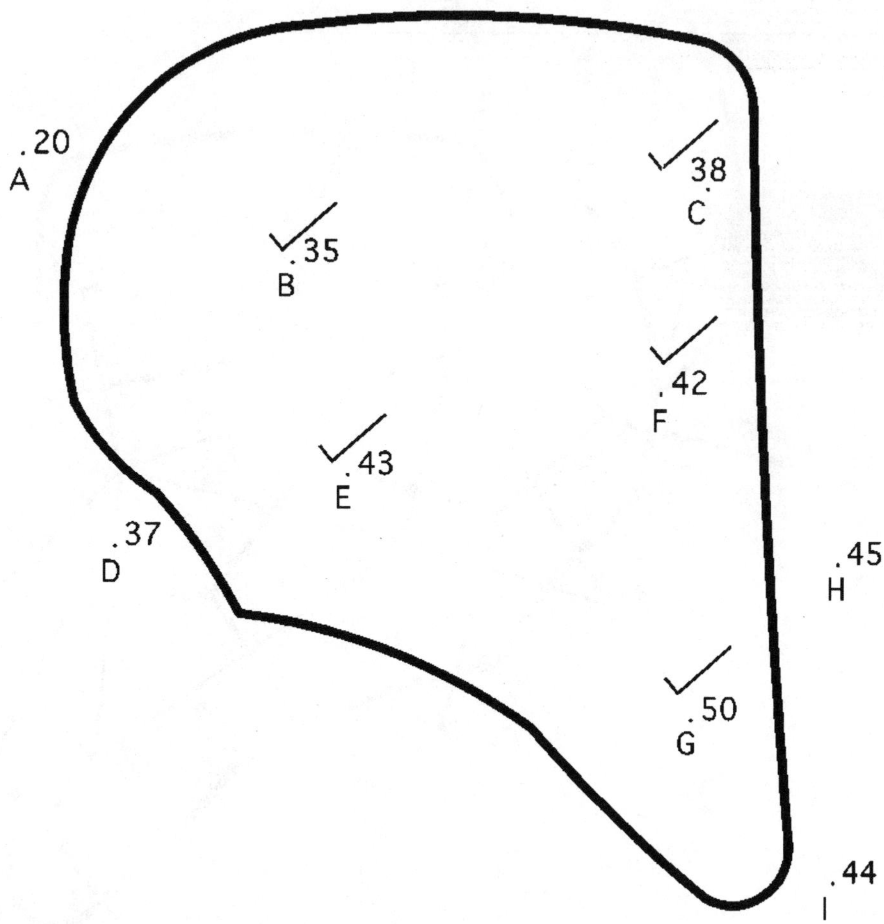

FIGURE E4.3.1b Using station average or arithmetic mean method.

Solution (2)

1 Station	2 Precip P	3 Square Area A	4 $P \times A$
A	20	31	620
B	35	79	2765
C	38	53	2014
D	37	22	814
E	43	37	1519
F	42	48	2016
G	50	51	2550
H	45	4	180
I	44	8	352
	$\sum$	333	12830

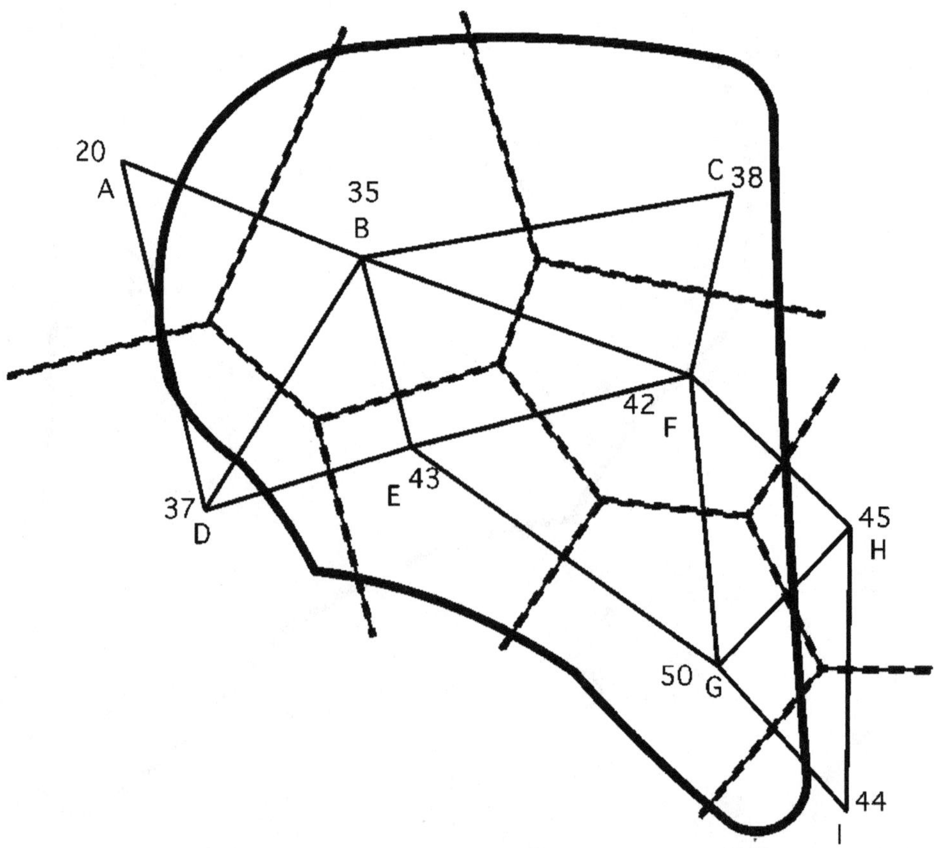

FIGURE E4.3.1c Thiessen polygon method.

Thiessen polygon method

$$\text{Average rainfall } \frac{\sum col\ (4)}{\sum col\ (3)} = \frac{\sum PiAi}{\sum Ai} = \frac{12830}{333} = 38.5 \text{ mm}$$

Solution (3)

1 Isohyet	2 Area A	3 $P_i \times A_i$
20	40	800
35	125	4375
40	100	4000
45	42	1890
50	26	1300
$\sum$	333	12365

$$\bar{P} = \frac{\sum col\ (3)}{\sum col\ (2)} = \frac{\sum PiAi}{\sum Ai} = 37.1 \text{ mm}$$

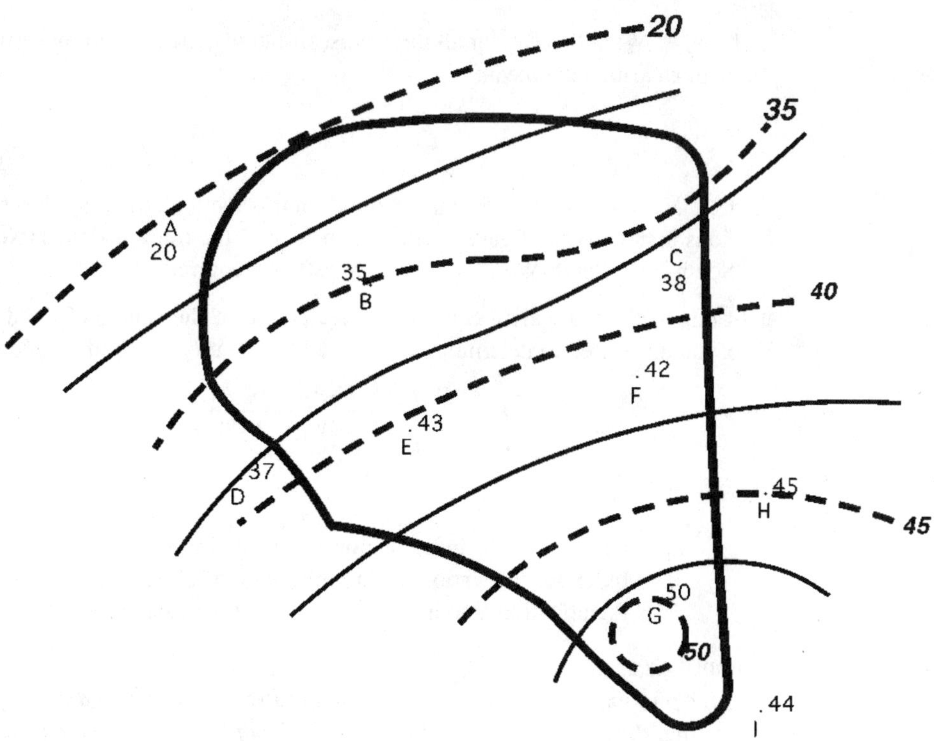

FIGURE E4.3.1d Isohyetal method.

4.3.4 Estimating Missing Rainfall Data

Sometimes a gage in a network of gages malfunctions or becomes temporarily inoperative during a storm. Whatever the case may be, it is important to be able to estimate the missing record for the particular storm or rainfall event so that a complete set of records is maintained for each rainfall gage.

Below, two methods for estimating the missing data are presented.

1. Normal ratio method. Suppose that the network of gages consists of n stations. These gage stations are referred to as the index stations.

The estimated rainfall for the gage x with missing data is:

$$P_x = \frac{N_x}{n}\left[\frac{P_1}{N_1} + \frac{P_2}{N_2} + \frac{P_3}{N_3} + \cdots + \frac{P_n}{N_n}\right]$$

In practice, it is sufficient to consider three closest index stations. Therefore:

$$P_x = \frac{N_x}{3}\left[\frac{P_1}{N_1} + \frac{P_2}{N_2} + \frac{P_3}{N_3}\right]$$

where N_i = Annual rainfall amount for gage station i (in the network of gages)
P_i = Precipitation amount recorded for the missing rainfall event for gage i
P_x = Missing rainfall amount to be estimated (that is, for gage x)

If $|N_x - N_i| \leq 0.10\,N_x$ for all three closest stations, then the normal ratio method simplifies to an arithmetic mean:

$$P_x = \frac{1}{3}[P_1 + P_2 + P_3]$$

2. U.S. National Weather Service method. In this method, the four closest index stations in a network of gages, with one in each of the four quadrants (North-East, South-East, South-West, and North-West) are required.

In this method, a weight equal to the reciprocal of the square of the distance of each index station is used to determine the estimate of the missing rainfall record.

$$P_x = \frac{W_1P_1 + W_2P_2 + W_3P_3 + W_4P_4}{W_1 + W_2 + W_3 + W_4}$$

where $W_i = \dfrac{1}{L_i^2}$
L_i = Distance of the index station i from station x
X = Index for the station having missing rainfall record
P_i = Precipitation amount for station i for the missing record.

EXAMPLE 4.3.2

The precipitation gage for station x was inoperative for a certain day in April, 2000. During that day, the rainfall record at six nearby index stations were collected and are given in the table below.

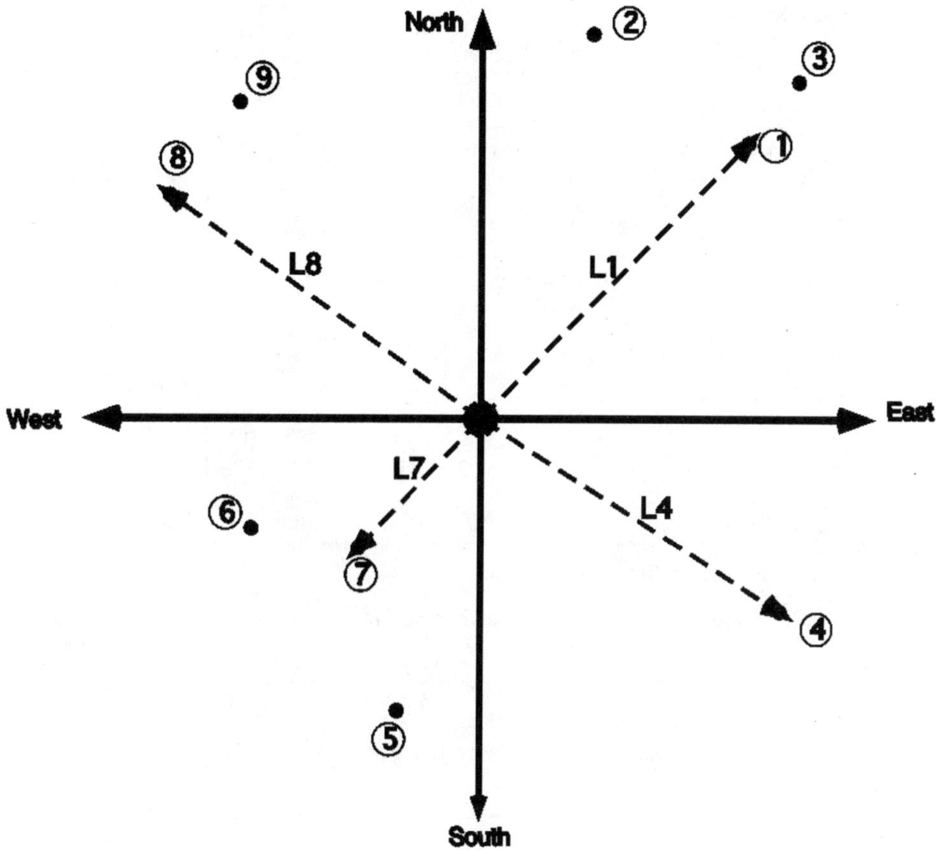

FIGURE 4.3.3 Quadrants for National Weather Service method.

1. *Determine the estimated rainfall amount for station x using the normal ratio method.*
2. *Estimate the missing rainfall record for station x using the U.S. National Weather Service method.*

Station	Quadrant	Distance from x, km	Rainfall, mm	Mean annual rainfall, mm
1	North-East	0.85	20	400
2	North-East	0.95	13	420
3	North-West	0.62	18	375
4	South-East	0.37	25	365
5	South-East	0.25	19	350
6	South-West	0.88	20	325
x		0	missing	285

Solution

1. The three closest stations are:

Station 5, Station 4, and Station 3

Checking $|N_x - N_i| \leq 0.10\, N_x$:

$$|N_x - N_5| = |285 - 350| = \frac{65}{285} = 0.228$$

Therefore must use normal ratio formula

$$P_x = \frac{N_x}{3}\left[\frac{P_5}{N_5} + \frac{P_4}{N_4} + \frac{P_3}{N_3}\right]$$

$$= \frac{285}{3}\left[\frac{19}{350} + \frac{25}{365} + \frac{18}{375}\right]$$

$$= 95(0.05429 + 0.06849 + 0.04800)$$

Missing estimate = 16.2 mm

2. Using the U.S National Weather Service method

$$P_x = \frac{W_1P_1 + W_2P_2 + W_3P_3}{W_1 + W_2 + W_3 + W_4}$$

The closest stations are (in each quadrant)

Station	Distance	$W_i = \frac{1}{L_i^2}$	P_i, mm
1	0.85	1.3841	20
3	0.62	2.6015	18
5	0.25	16.0000	19
6	0.88	1.2913	20

Sum $W_i = 21.2769$

$$P_x = \frac{W_1P_1 + W_3P_3 + W_5P_5 + W_6P_6}{\sum_{1,\,3,\,5,\,6} W_i}$$

$$= \frac{1.3841(20) + 2.6015(18) + 16.0(19) + 1.2913(20)}{1.3841 + 2.6015 + 16.00 + 1.2913}$$

$$= \frac{404.33}{21.277} = 19.0 \text{ mm}$$

Therefore, missing estimate is 19 mm.

4.3.5 Testing the Consistency of Gage Records

Several factors can affect the consistency of the rainfall data collected at a particular station. Changes in the procedure for data collection, exposure (vegetative cover for example because of de-treeing), personnel changes, gage relocation, or the installation of new instrumentation could lead to a change of a record's regime.

In hydrology, a method called *double-mass analysis* can be used to evaluate the records compiled at a gage station.

Double-mass analysis is used to test the consistency of a set of records by comparing (over the same period) annual or seasonal data compiled for the station x being tested, with similar data collected for a group of *reference stations* (usually in close proximity of the test station x. The comparison is accomplished by plotting cumulative annual or seasonal data, as shown in Figure 4.3.4.

In the figure, reference values are plotted on the abscissa and the test station values are plotted on the ordinate.

It can be seen that there is a break in the line as the slope changed starting in 1996. In interpreting the plot, one could say that: (1) the records at the test station are not consistent, (2) a change in regime began in 1996, and (3) to maintain the same regime, the data would have to be adjusted to match the regime before1996 or to match the regime after 1995.

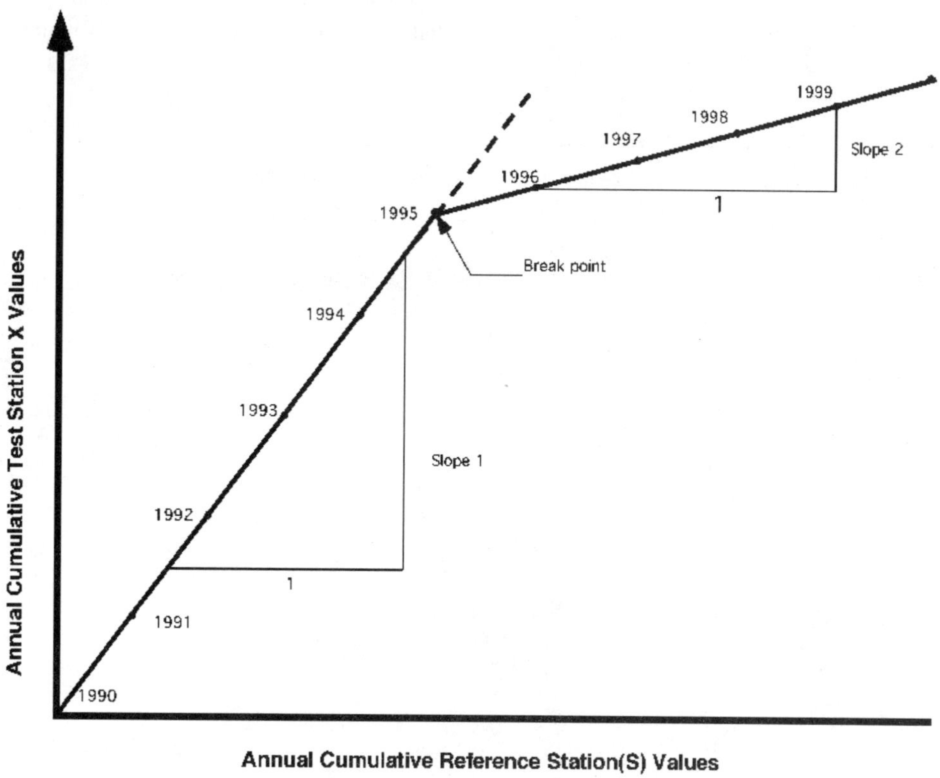

FIGURE 4.3.4 Gage consistency: double mass curve.

1. To adjust the test station x values so that records after 1995 match those *before* 1996, multiply annual (or raw) test station x values by $\frac{\text{slope 1}}{\text{slope 2}}$

$$\text{Therefore adjusted } x \text{ value (after 1995)} = (x \text{ value after 1995}) \times \frac{\text{slope 1}}{\text{slope 2}}$$

2. To adjust the values for station x so that the records before 1996 match the ones after 1995:

$$\text{multiply the annual (or raw) station } x \text{ values by } \frac{\text{slope 2}}{\text{slope 1}}$$

$$\text{Adjusted } x \text{ values (before 1996)} = (x \text{ value before 1996}) \times \frac{\text{slope 2}}{\text{slope 1}}$$

EXAMPLE 4.3.3

The annual precipitation data at rainfall gage station x on the University of Nigeria, Nsukka Campus, were to be tested for consistency from 1972 through 1987. Data at eight neighboring gage stations also were obtained and are presented on the following table.

1. Test the consistency of station x records. If a change in regime is identified, report the year in which the change occurred.
2. Adjust station x records so that the regime before the break is maintained for all values.

Year	Precipitation at x, mm	8-station average, mm
1972	35	28
1973	37	29
1974	39	31
1975	35	27
1976	30	25
1977	25	21
1978	20	17
1979	24	21
1980	30	26
1981	31	31
1982	35	36
1983	38	39
1984	40	44
1985	28	32
1986	25	30
1987	21	23

Solution

The table below presents the values for the double-mass curve. It is observed from the figure that a break in regime occurs after 1980. The adjusted values for the years 1981-1987 are shown on the table.

Year	Station x	Station Ref	Cumulative Values Cum ref	Cumulative Values cum x	Adjusted x values
1972	35	28	28	35	35
1973	37	29	57	72	37
1974	39	31	88	111	39
1975	35	27	115	145	35
1976	30	25	140	176	30
1977	25	21	161	201	25
1978	20	17	178	221	20
1979	24	21	199	245	24
1980	30	26	225	275	30
1981	31	31	256	306	40.8
1982	35	36	292	341	46.1
1983	38	39	331	379	50.1
1984	40	44	375	419	52.7
1985	28	22	407	447	38.9
1986	25	30	437	472	32.9
1987	21	23	460	493	27.7

4.4 THE HYDROLOGIC CYCLE AND THE HYDROLOGIC BUDGET EQUATION

The hydrologic cycle as has been defined is the continuous (perpetual) recirculatory path in which the waters of the earth circulate from the atmosphere to the earth's surface and underneath, perhaps reaching the groundwaters, and finally returning to the oceans where, through evaporation, the water returns to the atmosphere. The cycle of the water does not have to go through the entire system, as there is a continuous transfer from one phase to the other. The hydrologic cycle is driven by energy from the sun and the earth's gravitational field. The process is very dynamic.

In the hydrologic cycle, there are transport phases (liquid-transport and vapor-transport) and storage elements (vapor, liquid, and solid). Examples of these are listed in the following table.

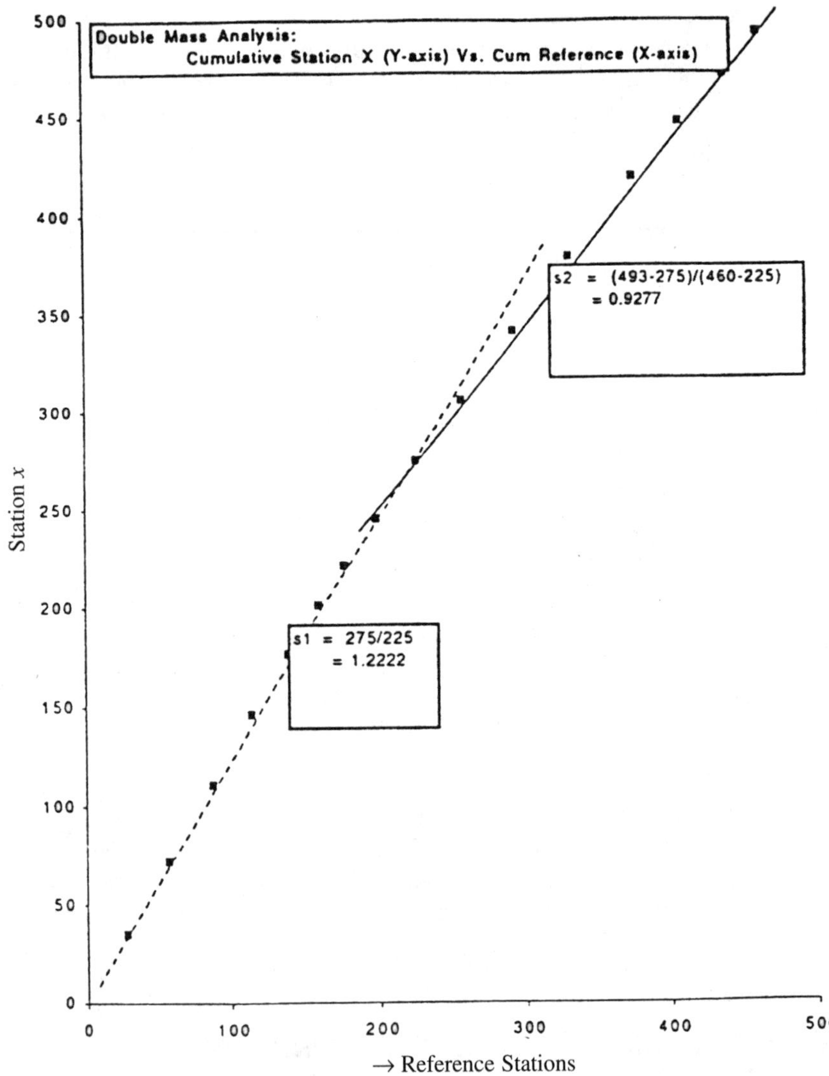

FIGURE E4.3.3

Liquid Transport Precipitation as Rainfall	Vapor Transport
Throughfall from cover to land surface	Evaporation from bodies of water to atmosphere
Percolation through soil to aquifers	Transpiration from plants to atmosphere
Interflow from soil to streams and rivers	Water vapor from soil
Surface runoff to rivers	Solid to vapor transfer from ice and snowcaps
Flow from rivers to oceans	
Snow melt on land surface	
Capillary rise in soil	

Water-holding Elements

Solid	Liquid	Vapor
Icecaps and snowpacks	Surface water (streams, rivers, ocean, lakes)	Atmosphere
Frozen soil	Groundwater	Soil
	Soil (porous media)	
	Plants	

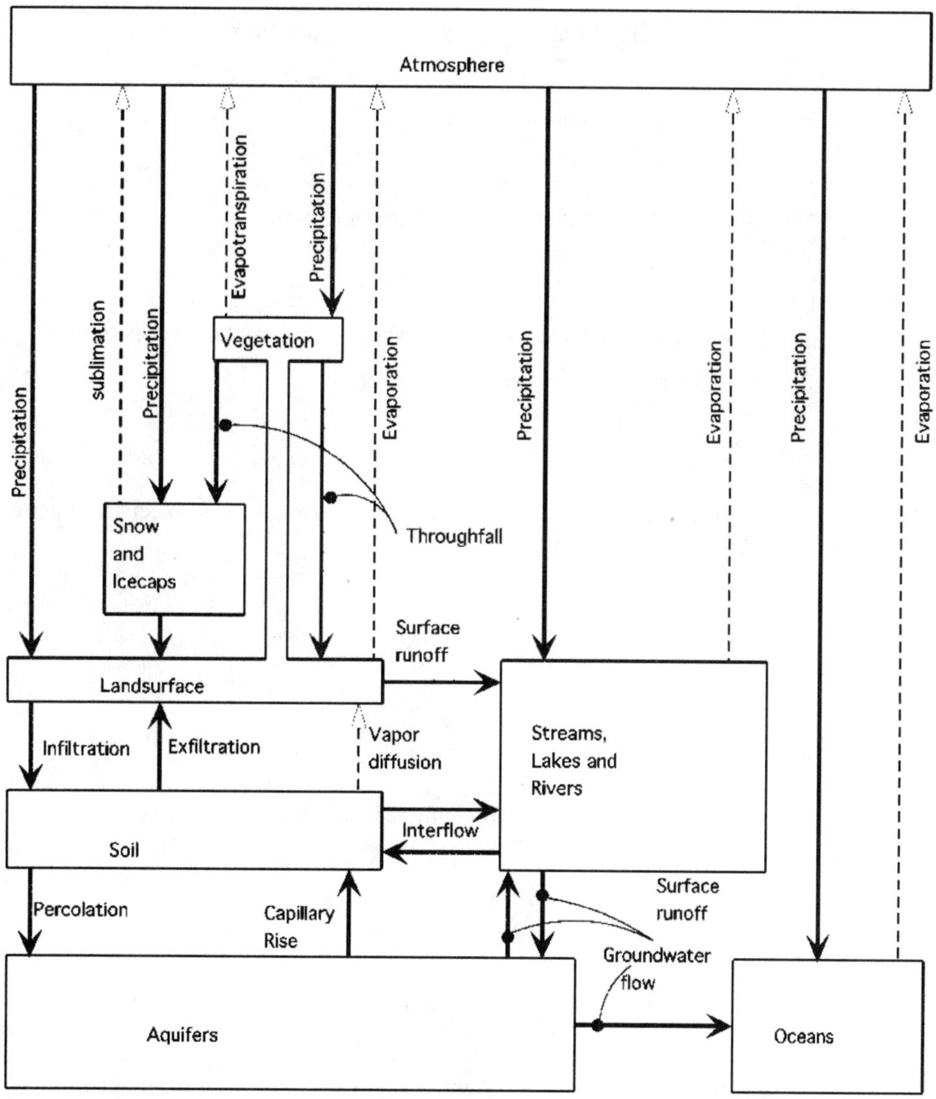

FIGURE 4.4.1 Hydrologic Cycle Schematic: Transport phases.

4.4.1 The Hydrologic Budget Equation

Recall that a watershed was defined as the portion of the earth's surface that collects runoff and concentrates it at a common point called the catchment outlet.

When the focus or area of interest is a watershed and the concept of the hydrologic cycle is described in that context, a relationship that accounts for the applicable components of the hydrologic for that watershed is called the hydrologic budget equation.

The hydrologic budget is an accounting of the various transport phases and storage elements of the hydrologic cycle within the boundaries of a watershed. In other words, it is a mass balance equation, in the context of the hydrologic cycle, for a given watershed.

There are two possible forms:

1. When surface water with groundwater interaction is considered:

$$\Delta s = P - (E + T + G + Q)$$

where, for the period in consideration,

P = Precipitation amount
E = Evaporation from the watershed
T = Transpiration from vegetation
G = Ground water flow amount
Q = Surface runoff volume.

2. When only surface water is considered and groundwater is ignored:

Change in storage cover a given period

$$\Delta s = P - (E + T + I + Q)$$

where I = Infiltration amount into the soil

In the last two equations, the terms in the parentheses, except for the runoff Q, often are combined and termed *losses*.

Thus

$$\Delta s = P - (\text{Loss} + Q)$$

Therefore

$$Q = P - \text{Loss} - \Delta s$$

In the special situation where there is no significant change in the storage for the watersheds:

$$\Delta s = 0,$$

so that

$$Q = P - \text{Loss}$$

The hydrologic budget equation can be expressed in volumetric units (m^3, ft^3, acre-ft), length units (mm, m, ft, in.) or volumetric flow rate units (m^3/s, ft^3/s, MGD, etc.)

It is understood that when length units are used, it is the average depth uniformly distributed over the entire watershed.

EXAMPLE 4.4.1

In a 600-hectare water supply watershed in the Piedmont Triad community of central North Carolina, the total precipitation over a period of 60 days was measured to be 208 mm. During the same period, inflow into and outflow out of the water supply lake was 2.05 m^3s and 1.85 m^3s, respectively. The seepage loss was 60 mm, and the loss to evaporation was 50 mm including transpiration from the vegetation.

1. Determine the change in storage in the watershed and give your answers in: a) m^3 and b) m^3s.
2. If the per capita daily water consumption is 150 liters, how many people will be served by an increase in storage of 4 mm/day in the watershed?

Solution

Area $= 600 \text{ hectares} = 600 \text{ ha} \times 10{,}000 \dfrac{\text{m}^2}{\text{ha}} = 6 \times 10^6 \text{ m}^2$

Time period $= 60 \text{ days} = 60 \times 24 \times 60 \times 60 \text{ s} = 5814000 \text{ s}$

Inflow Qi $= 2.05 \dfrac{\text{m}^3}{\text{s}} = 2.05 \dfrac{\text{m}^3}{\text{s}} \times 5184000 \text{ s} = 10627200 \text{ m}^3$

Outflow Qo $= 1.85 \dfrac{\text{m}^3}{\text{s}} = 1.85 \dfrac{\text{m}^3}{\text{s}} \times 5184000 \text{ s} = 9590400 \text{ m}^3$

Precipitation, P $= 108 \text{ mm} = 0.208 \text{ m}$

Precipitation volume $= \text{depth} \times \text{area} = 0.208 \text{ m} \times 6 \times 10^6 \text{ m}^2$

$= 1248000 \text{m}^3$

Seepage loss I $= 60 \text{ mm} = 0.060 \text{ m}$

Seepage volume $= \text{depth} \times \text{area} = 0.060 \times 6 \times 10^6$

$= 360000 \text{ m}^3$

Evaporation transpiration $ET = 50 = 0.050 \text{ m}$

Evapotranspiration volume $= \text{area} \times \text{depth} = 0.05 \text{ m} \times 6 \times 10^6 \text{ m}^2$

$= 300{,}000 \text{ m}^3$

Change in storage ΔS $= P - (ET + Q + I)$

But $Q = Qo - Qi$

Therefore $\Delta S = P - (ET + Qo - Qi + I)$

1. Change in storage
   a. Change in storage, m^3

$$\Delta S = 1{,}248{,}000 - (300{,}000 + 9{,}590{,}400 - 10{,}627{,}200 + 360{,}000)$$
$$= 1{,}248{,}000 - (-377{,}200) = 1{,}625{,}200 \text{ m}^3$$

Thus the change in storage was positive … more water.

b. To find the change in storage—volumetric rate of change, in this case m^3/s, divide the change by the time over which change occurred.

$$\Delta S\left[\frac{m^3}{s}\right] = \frac{1625200 \text{ m}^3}{5{,}184{,}000S} = 0.313 \frac{m^3}{s}$$

2. People served by increase in storage

$$\Delta S = 4 \frac{\text{mm}}{\text{day}} = 0.004 \times \text{Area} = 0.004 \text{ m} \times 6 \times 10^6 \text{ m}^2$$

$$= 24{,}000 \text{ m}^3 = 24{,}000 \text{ m}^3 \times 1000 \frac{1}{\text{m}^3} = 2.4 \times 10^7 \text{ l}$$

$$\text{daily per capita usage} = 150 \text{ l/person}$$

Therefore

$$\text{Number served} = \frac{2.4 \times 10^7 \text{ l}}{150 \text{ l/person}} = 160{,}000 \text{ persons}$$

4.5 HYDROLOGIC ABSTRACTIONS

When a watershed receives precipitation in any of its various forms, the resultant runoff is always less than the precipitation amount. Therefore, as stated in the last section, when there is no change in storage,

$$\text{Runoff, } Q = P - L$$

where L = Combined abstractive loss
P = Total precipitation.

Thus the abstractive loss L acts to reduce the precipitation to the resulting runoff. In fact, runoff is also called *effective precipitation,* and sometimes it is called *direct runoff.*

The processes that work together to reduce total precipitation to direct runoff are called *abstractive processes* or *hydrologic abstractions*, and they include: evaporation and transpiration, infiltration, interception, and surface or depression storage. Together, interception and depression storage are called *initial abstraction.*

Each of these abstractive processes will be discussed in detail in the sections that follow.

4.5.1 Interception

Interception, usually the first abstractive process during a storm, can be defined as the abstraction from precipitation due to vegetative or other (roofs, for example) surface cover. The part of precipitation that reaches the ground by first passing through vegetative cover is called *throughfall.* And *interception loss, L,* is the fraction of precipitation abstracted by interception. Interception loss depends on storm character (duration, intensity and type), the type and density of vegetative cover, and the time of the year or season.

Interception loss can be estimated by a relationship that includes a storage component and an evaporation component, thus:

$$\text{Interception loss, } L = S + KEt, \text{ mm}$$

where S = Interception storage that is the depth or volume retained by vegetative cover or other cover against wind and gravitational forces, mm
K = Factor that depends on the type of cover
E = Evaporation rate, mm\hr
t = Storm duration, hrs.

In urban areas, assuming about 10 percent of vegetative cover, about 2.5 mm can be intercepted during the first hour of a storm. In more densely vegetated areas, however, about 24 percent of the total annual precipitation is intercepted by trees.

Interception varies widely with the season of year. Some studies indicate a range of 7 percent to 36 percent during the growing season versus 3 percent to 22 percent during the dormant season. Furthermore, interception tends to be significant when the precipitation amount is low. In fact, interception loss over year can be up to 25 percent of the annual total and mostly from light storms.

Interception loss often is ignored in flood studies because the rainfall events that accompany flooding events are such that interception loss is an insignificant component.

4.5.2 Surface or Depression Storage

This is the process by which precipitation is abstracted (reducing runoff) by retention in surface depressions (for example, puddles, ditches, and any natural, man-made or artificial storage). The detained or stored precipitation evaporates into the atmosphere or infiltrates the ground.

Surface depressions are distributed randomly over a watershed, and thus the volume detained is difficult to estimate. The surface depression amount during a storm, however, is affected by the following:

1. Surface storage decreases with the slope of the watershed. The steeper the slope, the smaller the surface depression.
2. Surface storage is primarily satisfied during the earlier portion of a storm, becoming negligible as a storm lasts longer than 1 hour.

An estimate for surface or depression storage can be made with the following equation:

$$V_s = S_d(1 - e^{-kPe}), \text{ mm}$$

where S_d = Depression storage capacity, mm (range: 10-50 mm)
Pe = Precipitation excess, mm
$= P - L - I$
P = Total precipitation, mm
L = Interception loss, mm
I = Infiltration depth
K = A constant (assumed value = $1\backslash S_d$)

S_d for most watershed ranges from 10 mm to 50 mm and depends on the soils found in the watershed.

TABLE 4.5.1 Examples of Surface Depression Storage

Land Cover	Surface Storage, mm	Design Values, mm
Impervious areas		
Roofs, flat	1.3 – 3.8	2.5
Roofs, sloped	1.3 – 2.5	1.3
Pervious areas		
Lawn grass areas	5 – 13	8
Wooded areas and open fields	5 – 15	10

4.5.3 Infiltration

Infiltration is the process by which precipitation seeps into the soil below the land surface, thereby reducing runoff or direct precipitation.

Infiltration rates depend on the following factors:

1. Condition of the land surface and the moisture condition of the soil
2. Type, extent, and density of vegetation present
3. Physical properties of the soil including the type, grain size, and porosity and soil permeability
4. Time of year or season
5. Type of precipitation event: intensity, duration and amount of rainfall
6. Depth to the water table.

Infiltration can be described numerically by two methods—an instantaneous filtration method or an average infiltration method.

The instantaneous infiltration method describes infiltration by a continuous (smooth) relationship, which can be used to calculate the rate of infiltration at any instant. An example is given in Figure 4.5.1.

The second method, the average infiltration or the infiltration index approach, provides an average infiltration rate over the duration of the storm. This method is more practical for events that last a long time and for conditions where the watershed is already moist. In general, the assumption that a constant infiltration rate occurred over the entire storm tends to overestimate the infiltration rate at the latter part of a storm and underestimate the infiltration rate at the beginning of the storm, especially when dry conditions prevail at the beginning of a rainfall event.

4.5.4 Infiltration Rate Values

This section covers infiltration rates for several soil types and other general information that help to provide a better understanding of the role of infiltration. The objective is to

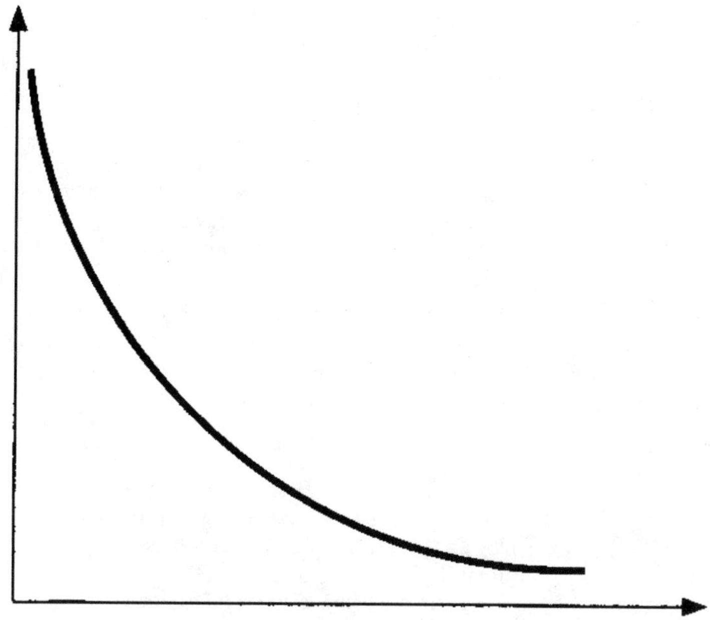

FIGURE 4.5.1 Typical instantaneous infiltration model.

provide the engineer with representative values and background information that may be very useful in practice.

Natural Resources Conservation Service, formerly Soil Conservation Service (SCS) of the U.S. Department of Agriculture, has provided a classification of soils into soil groups. That information is presented in Table 4.5.4.1.

A soils classification chart as provided by NRCS of USDA is provided below. The classification defines soil in terms of the percentage of sand, silt, and clay. These groups described by virtue of infiltration potential in Table 4.5.4.2 and the classification chart shown in Figure 4.5.2.

Table 4.5.4.3 gives the seasonal rainfall limits for antecedent moisture condition for watersheds. The table shows the three antecedent moisture conditions (AMC) that correspond to dry, average, and wet.

4.5.5 Instantaneous Infiltration Methods

1. Horton equation

$$f(t) = fc + (fa - fc)\mathbf{e}^{-Kt}$$

where $f(t)$ = Instantaneous infiltration rate at time t, mm, h, or other compatible units

fc = Final or ultimate infiltration rate, $\frac{\text{mm}}{\text{h}}$

fo = Initial infiltration rate, $\frac{\text{mm}}{\text{h}}$

K = Decay constant, h^{-1}

t = Time, h, or units compatible with K

TABLE 4.5.4.1 NRCS Soil Groups with Minimum Infiltration Rates

Hydrologic Soil Group	Description	Example Minimum Infiltration Rate, fc [mm\hr]
A	Lowest runoff potential and high infiltration rates. Includes deep sands with very little silt and clay; also, deep, rapidly permeable gravel. Have a high rate of water transmission through soil.	Sand = 210.1 Loamy Sand = 61.2
B	Moderately low runoff potential. Mostly sandy soils less deep and less aggregated than A, but the group as a whole has above average infiltration after thorough wetting. Moderate infiltration rate and moderate water transmission rate.	Sandy Loam = 25.9 Loam = 13.2
C	Moderately high runoff potential. Comprises shallow soils and soils containing considerable clay and colloids, though less than those of group D. The group has below-average infiltraion after saturation.	Silt Loam = 6.9 Sandy Clay Loam = 4.3
D	Highest runoff potential and lowest infiltration. Includes mostly clays of high swelling percentage, but the group also includes some shallow soils with nearly impermeable sub-horizons near the surface.	Clay Loam = 2.3 Silty Clay Loam = 1.5 Sandy Clay = 1.3 Clay = 0.5

Source: U.S. Department of Agriculture, Soil Survey Manual #18, Washington DC, 1951, and Wanielista, Kersten & English (1997), Hydrology Water Quantity and Quality Control, Wiley & Sons.

TABLE 4.5.4.2 AMC Classifications*

AMC	Total five-day Antecedent Rainfall, 9 mm	
	Dormant Season	Growing Season
I. A condition where soils are dry but not to the wilting point. Assuming satisfactory plowing or cultivation is practiced.	Less than 13 m	Less than 36 mm
II. Average case.	13 to 28 mm	36-53 mm
III. Wet situation. Heavy rainfall or light rainfall has occurred at low temperatures.	more than 28 mm	greater than 53 mm

* Data used to develop this table are from Midwestern United States. An adjustment may be necessary to account for local conditions.

When time $= o, f(o) = fo$
and after a long period has elapsed ($t < 10$ h), $t = \alpha\,; f(\alpha) = fc$

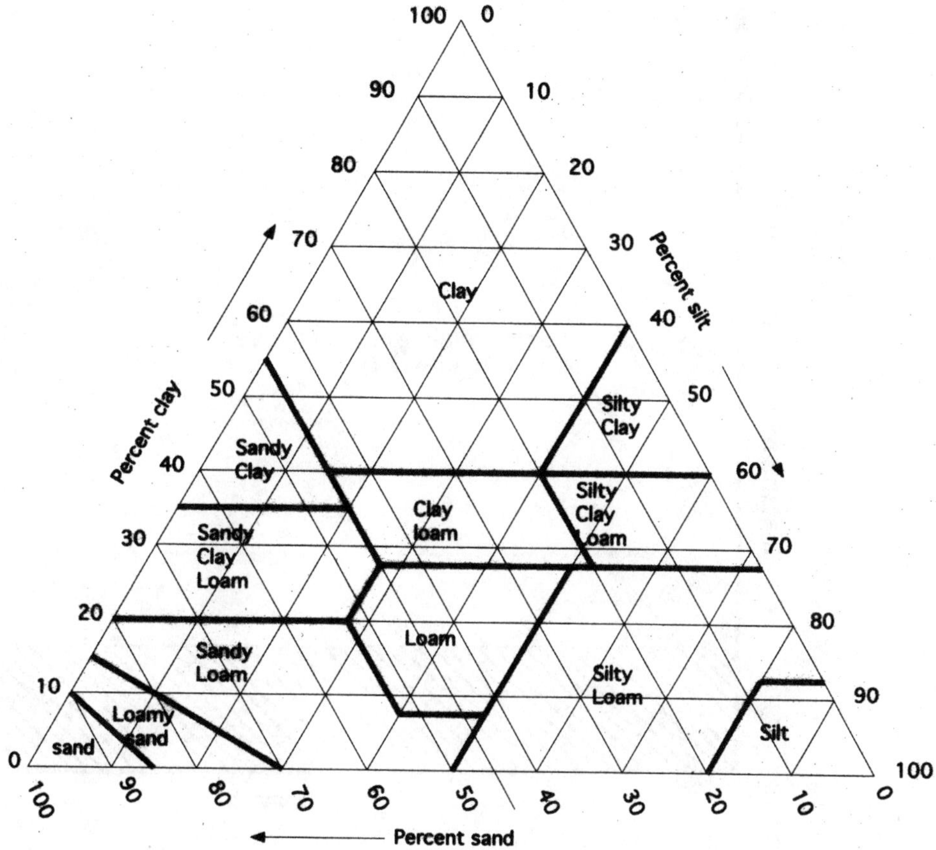

FIGURE 4.5.2 Textural soil classification triangle. *(USDA, 1951)*

To calculate total infiltration F

$$F = F_1 + F_2$$

$$F_1 = \int_o^{\alpha} (fo - fc)e^{-Kt} = \frac{fo - fc}{K}$$

$$F = F_1 + F_2 = fct + \left(\frac{fo - fc}{K}\right)$$

2. Philip's equation
In Philip's equation, the instantaneous infiltration rate is

$$f(t) = \tfrac{1}{2}St^{1\backslash 2} + A$$

where $f(t)$ = Instantaneous infiltration rate, mm\h
S = Empirical parameter indicating the rate of penetration
A = Parameter depending on soil
t = Time, hrs, or other compatible unit.

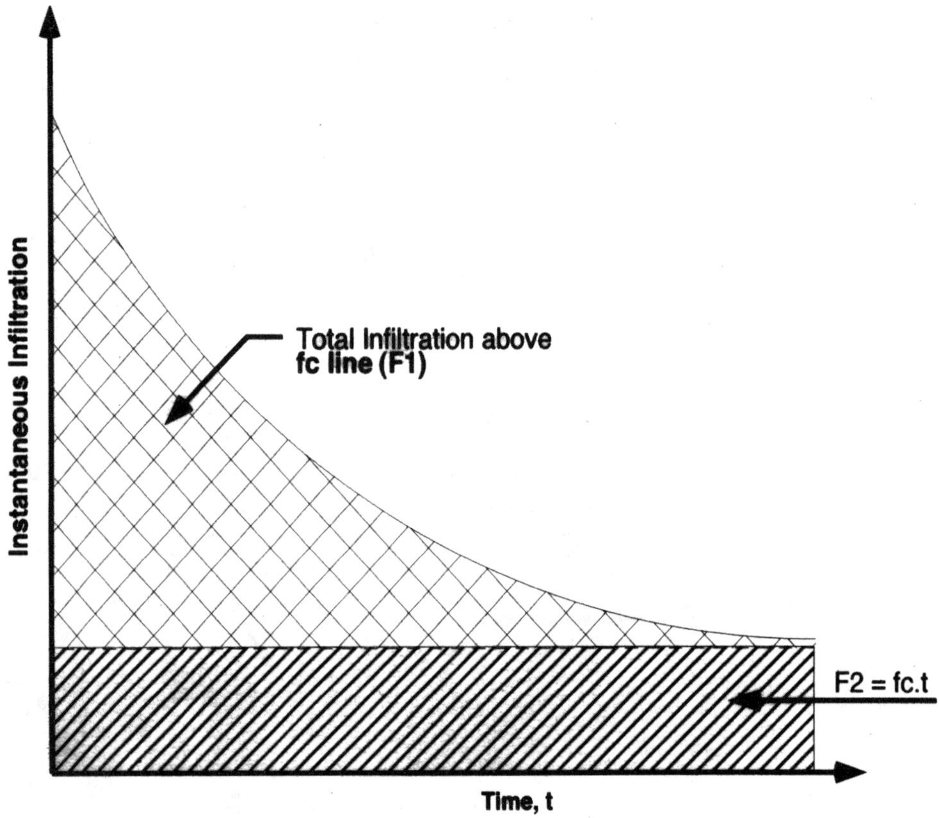

FIGURE 4.5.3 Total infiltration using Horton's equation.

A problem with Philip's equation is that infiltration is infinity at time = *o*. However, it has been shown that Philip's equation is still useful although this anomaly exists.

Total infiltration,

$$F = \int_o^t \left(\frac{1}{2} St^{-1\backslash 2} + A\right) dt$$
$$= St^{1\backslash 2} + At$$

EXAMPLE 4.5.1 APPLICATION OF HORTON'S EQUATION

The Horton's equation parameters for a certain watershed in West Lafayette, IN, are as follows:

$$f_o = 3.5 \frac{\text{mm}}{\text{hr}}, fc = 0.50 \frac{\text{mm}}{\text{hr}} \text{ and the constant } K = 0.90 \text{ h}^{-1}$$

1. *Provide the instantaneous infiltration, f versus time, h, curve at the times, h: 0.0, 0.10, 0.20, 0.40, 0.80, 1.00, 2.00, 3.00, 4.00, 5.00, 6.00, 7.00, 8.00, 9.00, 10.00, 15.00, 20.00.*

2. *Calculate the total infiltration, mm, over the period of 20 h.*

Solution

1.

Time (hrs)	Capacity, f (mm/hr)
00.00	03.500
00.10	03.242
00.20	03.006
00.40	02.593
00.80	01.960
01.00	01.720
02.00	00.996
03.00	00.702
04.00	00.582
05.00	00.533
06.00	00.514
07.00	00.506
08.00	00.502
09.00	00.501
10.00	00.500
15.00	00.500
20.00	00.500

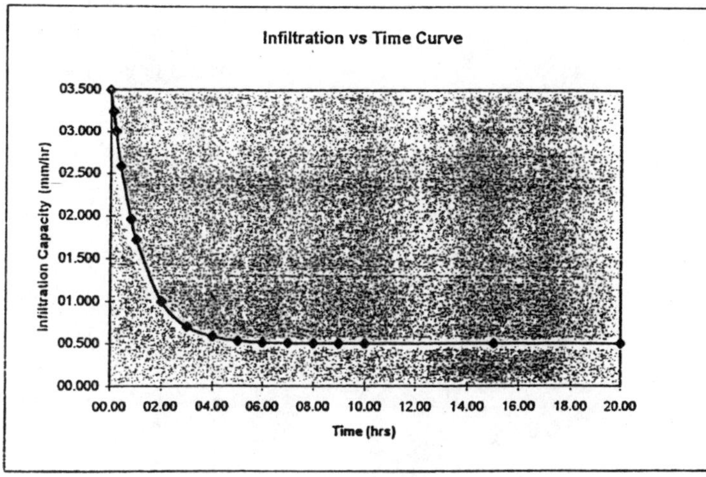

FIGURE E4.7.1A

2. Total infiltration,

$$F = f_c t + fo - \frac{fc}{K}$$

$$= 0.50\,\frac{\text{mm}}{\text{h}}\,[20\,h] + \frac{3.50\,\frac{\text{mm}}{\text{h}} - 0.50\,\frac{\text{mm}}{\text{h}}}{0.90\,h^{-1}}$$

$$= 10 + 3.33$$

$$= 13.33\text{ mm}$$

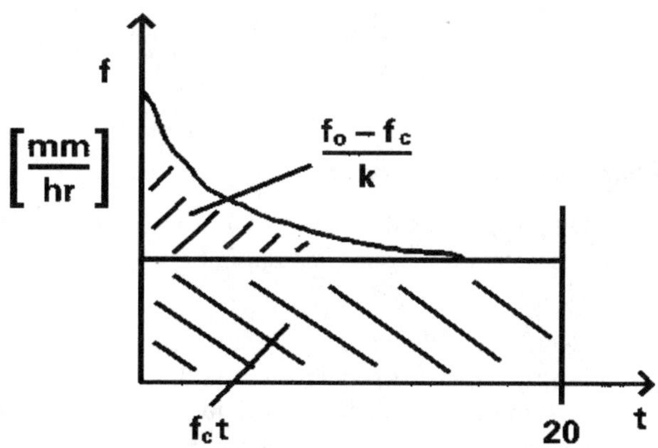

FIGURE E4.7.1B

EXAMPLE 4.5.2 APPLICATION OF PHILIP'S EQUATION

Determine the parameters S and A for the Philip's infiltration model given the following field measurements for instantaneous infiltration values f

$$\text{At 2.1 hours, } f = 1.9\ \frac{\text{mm}}{\text{hr}}$$

$$\text{At 5.3 hours, } f = 1.5\ \frac{\text{mm}}{\text{hr}}$$

Solution

Philip's equation is given by:

$$F = \frac{0.5S}{\sqrt{t} + A}$$

Therefore at time = 2.1 hours:

$$1.9 = \frac{0.5S}{\sqrt{2.1}} + A = 0.345S + A$$

At time = 5.3 hours:

$$1.5 = \frac{0.5S}{\sqrt{5.3}} + A = 0.2172S + A$$

Combining the two equations by subtracting the first from the second:

$$\begin{array}{r} 1.9 = 0.345S + A \\ -\ 1.5 = 0.2172S + A \\ \hline 0.4 = 0.1278S \end{array}$$

Therefore $S = 3.13$

From the first equation: $1.9 = 0.345(3.13) + A$

Therefore $A = 0.82$

Therefore, the Philip's model is: $f = \dfrac{1.565}{\sqrt{t} + A} + 0.82$

4.5.4 Infiltration Indices—Average Infiltration Rate

Two infiltration indices are presented in this section. Discussed above, the indices assume that the infiltration rate is constant throughout the rainfall event.

4.5.4.1 ϕ-Index. The ϕ-index is defined as a constant infiltration rate to be subtracted from the prevailing rainfall rate in order to obtain the runoff volume that was observed. In this method, it is always best to first provide the rainfall hyetograph. Next, a constant infiltration rate *range* is assumed. Based on the assumption, the ϕ-index is calculated so that the volume of water above the assumed ϕ-line is equal to the runoff volume given. If the calculated ϕ value is in agreement with the assumed ϕ-value range, then the correct ϕ has been found, otherwise another ϕ range is assumed and the calculations are repeated until a value of ϕ that is compatible with assumptions made is found.

Mathematically,

$$\phi\text{–}index = \frac{P_\phi - Q}{t_\phi} \frac{\text{mm}}{\text{hr}}$$

where P_ϕ = Total precipitation amount, mm, when the rainfall intensity is assumed to be greater than the Φ-index value assumed

t_ϕ = Total duration, h, when the rainfall intensity is assumed to be greater than the assumed Φ value

Q = Given amount of the total runoff from the watershed during the rainfall event, mm.

EXAMPLE 4.5.3

For the following temporal rainfall distribution, the runoff depth has been estimated to be 36 mm, and the sum of depression storage loss and interception is estimated to be 20 mm.

Calculate the ϕ-index. Begin by providing a sketch of the rainfall hyetograph. Show your work fully.

Solution

Time, $h \Rightarrow$	0–2	2–5	5–6	6–9	9–10	10–12
Rainfall intensity, mm/hr ⇒	6	15	12	8	10	5

Solution

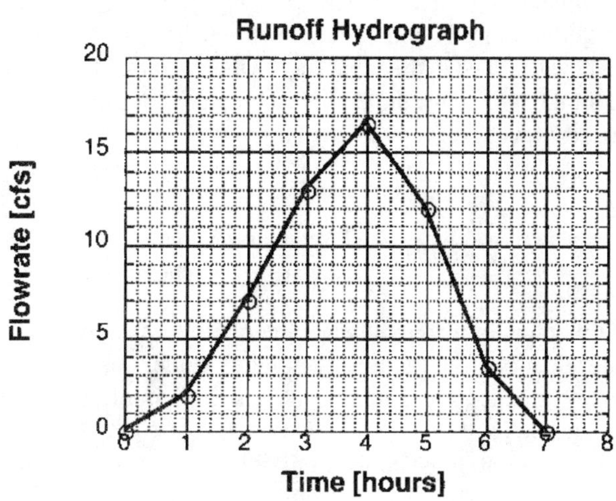

FIGURE E4.5.3

Trial #1

Assume that $5 \leq \phi \leq 6$

$$\phi = \frac{P_\phi - Q}{t_\phi}$$

$$P_\Phi = 6 \times 2 + 15 \times 3 + 12 \times 1 + 8 \times 3 + 10 \times 1 = 103 \text{ mm}$$

$$t_\phi = 2 + 3 + 1 + 3 + 1 = 10 \text{ hr}$$

$$\phi = \frac{103 - 36}{10} = 6.7 \frac{\text{mm}}{\text{hr}} \quad \text{(Not OK)}$$

The ϕ value does not agree with the assumption.

Trial #2

Assume that,

$$6 \leq \phi \leq 8$$

$$P_\phi = 15 \times 3 + 12 \times 1 + 8 \times 3 + 10 \times 1 = 91 \text{ mm}$$

$$t_\phi = 3 + 1 + 3 + 1 = 8 \text{ hr}$$

$$\phi = \frac{91 - 36}{8} = 6.9 \frac{\text{mm}}{\text{hr}} \quad \text{(good)}$$

This satisfies the assumption.

4.5.4.2 W-index The W-index is an average infiltration index similar to the Φ-index, however, it considers initial abstraction (interception and depression storage) while the Φ-index does not. Under extremely wet conditions, the *W*-index and the Φ-index would give approximately the same results because initial abstraction will be virtually nil. Furthermore, for storms with very long duration, *W*- and ϕ-index will be virtually the same.

The *W*-index can be calculated in the same way as the ϕ-index (described in the previous section) except that in the numerator the initial abstractions (depression storage and interception) are subtracted as well as the runoff volume.

Mathematically,

$$W - index = \frac{P_w - Q - IA}{t_w} \frac{\text{mm}}{\text{hr}}$$

where P_w = Total precipitation amount when the rainfall intensity is assumed to be greater than the trial (assumed) *W*-index value, mm

IA = Initial abstraction amount, mm. This is the sum of depression storage and interception loss

t_w = Total time duration when the rainfall intensity is greater than the trial *W*-index value, h.

EXAMPLE 4.5.4

For the following temporal rainfall distribution, the runoff depth has been estimated to be 36 mm and the sum of depression storage loss and interception is estimated to be 20 mm.

Calculate the W-index

Begin by providing a sketch of the rainfall hyetograph. Show your work fully.

Solution

Time, $h \Rightarrow$	0−2	2−5	5−6	6−9	9−10	10−12
Rainfall intensity, mm/hr $\Rightarrow$	6	15	12	8	10	5

Trial #1

Assume that,

$$6 \le W \le 8$$

$$W = \frac{P_w - Q - IA}{t_w}$$

where $IA = 20$ mm;

$Q = 36$ mm

$P_w = 15 \times 3 + 12 \times 1 + 8 \times 3 + 10 \times 1 = 91$ mm

$t_w = 3 + 1 + 3 + 1 = 8$ hr

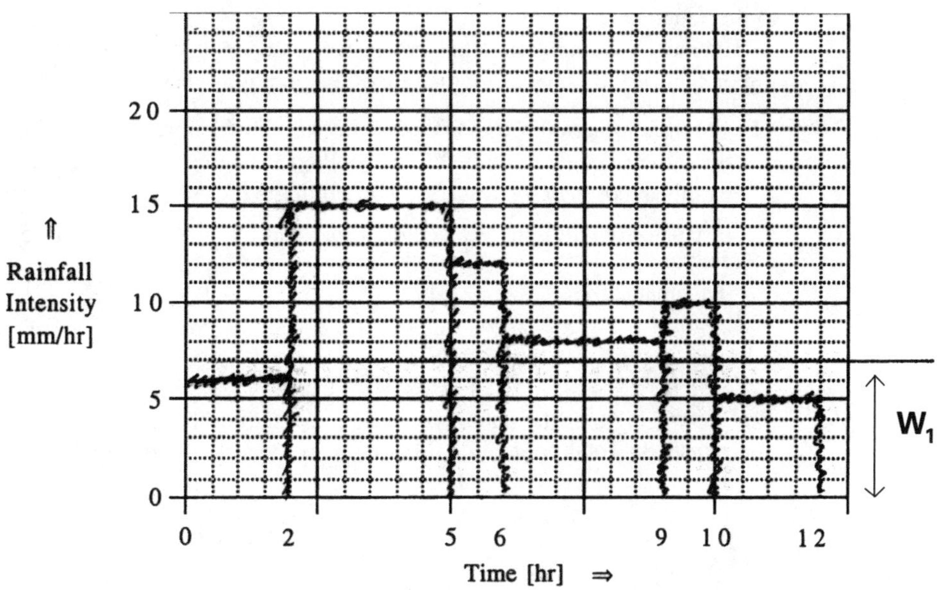

FIGURE E4.5.4A

Therefore,

$$W = \frac{91 - 36 - 10}{8} = 4.4 \frac{\text{mm}}{\text{hr}} \quad \text{(not in range 6–8)}$$

Since the value 4.4 $\frac{\text{mm}}{\text{hr}}$ is not within the assumed range of 6–8 $\frac{\text{mm}}{\text{hr}}$, the assumption was not satisfied. A new trial is necessary.

Trial #2

Assume that $0 \le W \le 5$

$$P_w = 6 \times 2 + 15 \times 3 + 12 \times 1 + 8 \times 3 + 10 \times 1 + 5 \times 2 = 113 \text{ mm}$$

$$t_w = 2 + 3 + 1 + 3 + 1 + 2 = 12 \text{ hr}$$

$$W = \frac{P_w - Q - IA}{t_w} = \frac{113 - 36 - 20}{12} = 4.8 \frac{\text{mm}}{\text{hr}} \quad \text{(good)}$$

This answer satisfies the assumption that W is between 0–5 $\frac{\text{mm}}{\text{hr}}$.

4.5.5 Evaporation

Evaporation is the conversion of water on the land surface, including surface depression storage lakes, reservoirs, streams, rivers, surface cover, snow caps, and snowpacks to water vapor. Because there is usually a continuous exchange of (liquid) water molecules into water vapor and vice versa (at the liquid air interface), evaporation is a dynamic process, and therefore evaporation refers to net evaporation—the net rate of conversion of water (liquid or solid) to the vapor state.

The rate of evaporation depends on several factors that can combine to result in limiting or significantly increasing the rate of evaporation. These factors include:

1. Net solar radiation—the difference between incident radiation and reflected radiation. The *solar constant* is the solar radiation that reaches the upper limits of the earth's atmosphere on a surface normal to the incident radiation at the earth's mean distance from the sun. The solar constant has a range of 1.89 to 2.05 = $\frac{\text{ly}}{\text{min}}$.

$$1 \text{ ly} = 1 \text{ Langley} = 1 \frac{\text{cal}}{\text{cm}^2}$$

The average solar constant is approximately 2 $\frac{\text{ly}}{\text{min}}$ (actually 1.97 cal\min).

Albedo (A) is the ratio (given in percentage) of the visible radiation that is reflected to the visible radiation that is incident on a surface. Thus:

$$\text{Albedo } A = \frac{\text{reflected sunlight}}{\text{incident sunlight}}$$

Albedo varies with location on the earth. It is less at high solar altitude and depends on the type of surface or cover. Generally albedo is less on moist surface than for dry surfaces. Since water or ice molecules require a certain amount of heat (equal to the latent heat) to be excited to effect a phase change, the more heat (due to radiation) that is available due to solar radiation, the higher the rate of evaporation.

2. Evaporation rate is affected by the saturation vapor pressure. Because the vapor pressure of a liquid depends on its temperature as the vapor pressure increases, the evaporation rate will tend to increase as liquid molecules because of the higher pressure gradient over the atmosphere (or ambient pressure). Evaporation will proceed at a higher rate until the pressure gradient is equalized.
3. Wind velocity affects evaporation by generating convective transport processes that transport water vapor (at the evaporating surface). Thus, the higher the prevailing wind speed the greater the evaporation rate.
4. The temperature of air and the water affect evaporation by increasing vapor pressure and supplying the latent heat of vaporization.

All of these factors reflect the weather at a particular location and, in the long run, the climate. Therefore, it is evident that the evaporation rate of a locale depends on the climate. Thus, the dependence of evaporation rates on climatic factors is taken into account when developing large-scale water resources projects that involve surface water. In arid or semiarid areas, a detail evaluation of potential evaporation is necessary in the planning and design of storage reservoirs.

4.5.6 Estimation of Evaporation Rates

Several techniques have been devised to calculate evaporation from bodies of water because it is difficult to measure evaporation directly. These methods are reviewed briefly here, and where appropriate, the models for the numerical evaluation of evaporation rates are presented.

The following methods may be used:

1. Water budget method
2. Energy budget method
3. Mass transfer approach
4. Combination methods
5. Pan evaporation/pan coefficient method.

4.5.6.1 Water Budget Method. In this method, the hydrologic budget equation is used to calculate the evaporation amount or the evaporation rate. The assumption made is that all the other terms in the equation can be determined, leaving the evaporation term as the subject of the equation. Therefore,

$$E = P + Q - O - GW - \Delta S, \text{ mm}$$

This method is theoretically simple, but it is hampered by the difficulties encountered in estimating the groundwater (or seepage) term. Other problems involve the accuracy of precipitation, inflow, outflow, and change in storage estimates.

In the above equation (for a given period),

P = Precipitation received in watershed, mm

Q = Surface runoff (inflow into lake or reservoir) in watershed, mm

O = Outflow or discharge (from lake or reservoir)

GW = Seepage or infiltration or groundwater flow out of watershed, mm

ΔS = Change in storage in the watershed, mm.

EXAMPLE 4.5.5

A lake with an average surface area of 360 ha (hectare) experiences a water level drop of 50 mm in a period of 25 days. The rainfall during the same period was 100 mm. Inflow and outflow from the reservoir were 0.25 $\frac{m^3}{s}$ *and 0.12* $\frac{m^3}{s}$ *respectively. Although seepage loss is small, it amounts to an estimated 3 mm every 10 days. Estimate the evaporation rate from the reservoir in* $\frac{mm}{day}$.

Solution

For a period of 1 day:

$$E = P + Q - O - GW - \Delta S \ \frac{\text{mm}}{\text{day}}$$

(These quantities are as previously defined)

$$P \text{ (precipitation)} = \frac{100 \text{ mm}}{25 \text{ day}} = 4.0 \ \frac{\text{mm}}{\text{day}}$$

$$Q \text{ (inflow)} = 0.25 \ \frac{\text{m}^3}{\text{s}} = 0.25 \times 10^6 \ \frac{\text{cm}^3}{\text{s}} \times 86400 \ \frac{S}{\text{day}}$$

$$= 2.16 \times 10^{10} \ \frac{\text{cm}^3}{\text{day}}$$

$$Q\left(\frac{\text{cm}}{\text{day}}\right) = \frac{\frac{\text{Vol}}{\text{day}}}{\text{Area of lake, cm}^2} = \frac{2.16 \times 10^{10} \ \frac{\text{cm}^3}{\text{day}}}{360 \text{ ha} \times 10{,}000 \ \frac{\text{m}^2}{\text{ha}} \times 10^4 \ \frac{\text{cm}^2}{\text{m}^2}}$$

$$Q = 0.60 \ \frac{\text{cm}}{\text{day}}$$

$$Q = 6.0 \ \frac{\text{mm}}{\text{day}}$$

Similarly, Outflow,

$$O\left(\frac{\text{cm}^3}{\text{day}}\right) = 0.12 \times 10^6 \ \frac{\text{cm}^3}{S} \times 86400$$

$$= \frac{1.0368 \times 10^{10}}{360 \times 10^8} = 0.288 \ \frac{\text{cm}}{\text{day}}$$

$$O = 2.9\ \frac{\text{mm}}{\text{day}}$$

$$\text{GW} = (\text{seepage}) = \frac{3\ \text{mm}}{10\ \text{days}} = 0.3\ \frac{\text{mm}}{\text{day}}$$

$$\text{Change in storage} = -\frac{50\ \text{mm}}{25\ \text{days}} = -2\ \frac{\text{mm}}{\text{day}}$$

$$E\left(\frac{\text{mm}}{\text{day}}\right) = 4.0 + 6.0 - 2.9 - 0.3 - (-2)$$

$$= 8.8\ \frac{\text{mm}}{\text{day}}$$

4.5.6.2 Energy Budget Method The energy budget method considers the conservation of energy over a given period of time rather than the conservation of mass (the water budget approach).

In this method, the energy available to effect evaporation is calculated (estimated) by combining the estimates of the outgoing energy, incoming energy, and the energy stored in the body of water (lake or reservoir usually) over a given period of time.

In order to fully understand the procedure for estimating evaporation by this approach, additional complementary information is provided in this section.

The latent heat of vaporization *Le* (evaporation) is the heat required to convert water to water vapor at the same temperature. The latent head of evaporation is equal to the latent heat of condensation *Lc*.

The latent heat of sublimation *Ls* is the heat required to convert ice to water vapor. And finally, the latent heat of melting *Lm* is the heat released when ice melts to water (at the same temperature).

The following equations can be used to estimate each of these quantities:

The latent heat of vaporization,

$$Le\left(\frac{\text{cal}}{\text{g}}\right) = 597.3 - 0.57T = -L_c$$

The latent heat of sublimation,

$$Ls\left(\frac{\text{cal}}{\text{g}}\right) = 677 - 0.07T$$

The latent heat of melting,

$$Lm\left(\frac{\text{cal}}{\text{g}}\right) = 79.7$$

where T is the temperature in degrees Celsius.

The heat flux $Qe\left(\frac{\text{cal}}{\text{cm}^2\text{-day}}\right)$ required to effect an evaporation rate $E\left(\frac{\text{cm}}{\text{day}}\right)$ at a certain temperature T is given by:

$$Qe = \rho LeE$$

where ρ = Density of water at temperature T, $\frac{\text{g}}{\text{cm}^3}$

Le = Latent heat of evaporation (vaporization) at temperature T, $\frac{\text{cal}}{\text{g}}$.

Therefore, if the net heat transfer rate (flux) to a body of water is known, the corresponding rate of evaporation, which will complete the energy balance situation, can be calculated.

Therefore,

$$E = \frac{Qe}{PLe}, \frac{\text{cm}}{\text{day}}$$

To use the energy balance method to calculate or estimate the evaporation rate from a body of water, a careful accounting of all the energy terms in the watershed must be done. Obviously, the accuracy of the accounting determines the accuracy of the evaporation rate estimates.

A significant and important component of the energy budget in a catchment is solar energy flux—more accurately the net energy flux due to the various forms of radiation. This includes net radiation from the sun and black body radiation from the body of water.

The net total radiation absorbed by a body of water (or snow pack) includes incident and reflected shortwave radiation (from the sun) and incident and reflected longwave (black body) radiation from the atmosphere and the surface of the water.

Radiation flux is measured in $\frac{\text{ly}}{\text{min}} \left(1 \frac{\text{angley}}{\text{min}} = \frac{\text{cal}}{\text{cm}^2\text{-min}}\right)$. In true SI units, $1 \frac{\text{ly}}{\text{min}} = 697.33 \frac{W}{\text{m}^2}$.

Incoming solar radiation (shortwave radiation) is called *insolation* (given traditionally in $\frac{\text{ly}}{\text{min}}$), and the incident radiation on a body at the base of the earth's atmosphere depends on the solar attitude, the normalized sun-earth radius (which depends on the particular location and the date), and other factors, including: cloud density, the relative path length of the sun's rays through the atmosphere (optical air mass), and an empirical transmission coefficient.

The reflected shortwave radiation from a body of water or snow pack is accounted for by a factor called the average albedo A, %. An average value of 7 percent can be used for water and for snow packs: 83 percent for new snow. For old snow, the range is from 42 to 70 percent. To calculate reflected shortwave radiation,

$$Q_{ir} = \frac{A \times Q_I}{100} \left(\frac{\text{cal}}{\text{cm}^2\text{-day}}\right)$$

where A = Albedo, %

Q_i = insolation, $\frac{\text{ly}}{\text{min}}$ or $\frac{\text{ly}}{\text{day}}$

Note that shortwave radiation only occurs during daylight hours.

Longwave radiation is due to the temperature of an object—hence, it is called *black body radiation*. All objects at temperature greater than 0 K (absolute zero temperature) emit radiation. The longwave atmospheric radiation is of interest here. This is called *atmospheric longwave radiation* Q_{atm}.

Just as shortwave radiation is reflected, so is longwave radiation. The accepted average albedo for atmospheric radiation is 3 percent.

Therefore,

$$Q_{\text{atmr}} = 0.03Q_{\text{atm}}$$

Back radiation is the name given to black body radiation from the water surface. Since black body radiation is dependent upon temperature, it can be estimated using the following temperature dependent relationship:

$$Q_{\text{back}} = 0.97\sigma T_s^4$$

where 0.97 is the assumed emissivity of 97%

σ = Stefan-Boltzman constant

$$\left(8.13 \times 10^{-11} \frac{\text{ly}}{\text{min} \cdot \text{K}}\right)$$

T_s = Water surface temperature in Kelvin (K)

Let

Qn = Net radiation absorbed by the body of water

Qi = Incident solar radiation or insolation

Therefore

$$Qn = Qi - Qir + Q_{\text{atm}} - Q_{\text{atmr}} - Q_{\text{back}}$$

If the albedo = A

$Qir = AQi$

Air = Reflected insolation

Further, assuming a reflectivity of 3 percent for atmospheric radiation,

$$\begin{aligned} Q_{\text{atm}} - Q_{\text{atmr}} &= Q_{\text{atm}}{}^{-0.03Q_{\text{atm}}} \\ &= 0.97Q_{\text{atm}} \end{aligned}$$

Therefore,

$$Qn = (1 - A)Qi + 0.97Q_{\text{atm}} - Q_{\text{back}}$$

In Figure 4.5.5, the following terms are defined thus:

Qn = Sensible heat transfer due to conduction and convection

Qe = Energy used for evaporation

Q_{stor} = Increase in energy store in the body of water

Q_{adv} = Energy change (increase) due to flow of water into and out of the lake or reservoir.

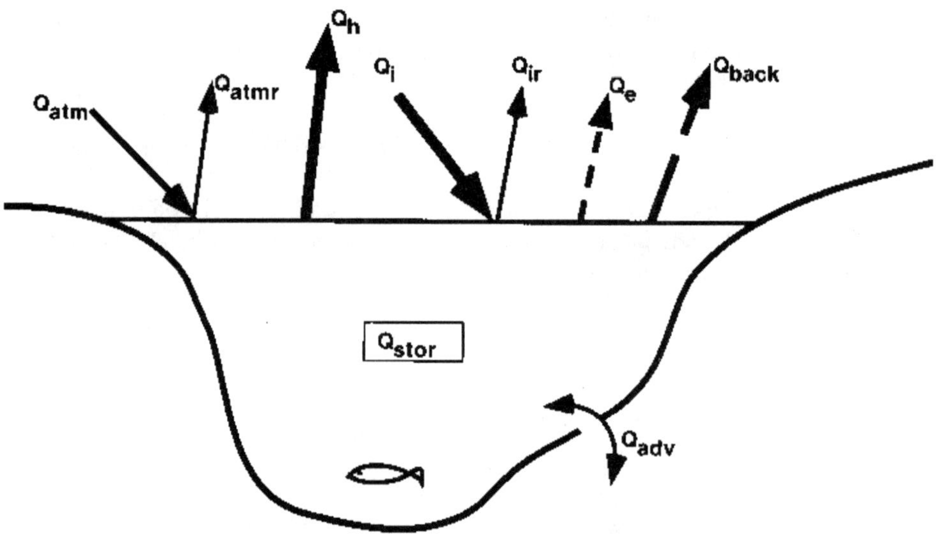

FIGURE 4.5.5 A schematic showing the energy budget scenario in a lake.

Therefore, energy change,

$$Q_{\text{stor}} = Q_n - Q_h - Q_e + Q_{\text{adv}}$$

Therefore, energy used for evaporation,

$$Q_e = Q_n - Q_h - Q_{\text{stor}} + Q_{\text{adv}}$$

But the evaporation rate,

$$E = \frac{Q_e}{\rho Le} = \frac{Q_n - Q_h - Q_{\text{stor}} + Q_{\text{adv}}}{\rho Le}$$

The ratio:

$$\frac{\text{heat loss by conduction/convection}}{\text{heat loss by evaporation}} = B$$

where B is the Bowen ratio, which is the measure of the sensible heat transfer, and is computed as

$$B = 0.66\left(\frac{T_s - T_a}{e_s - e_a}\right)\left(\frac{P}{1000}\right)$$

$$= \gamma\left(\frac{T_s - T_a}{e_s - e_a}\right)$$

where $\gamma = \dfrac{0.66P}{1000}$

B values are in the range 0.58–0.66 with a normal value of 0.61.

where P = Atmospheric pressure, millibars
e_s = Saturation vapor pressure of water at water temperature at the water surface
e_a = Vapor pressure of air, millibars
T_s = Water surface temperature, °C
T_a = Air temperature, °C

Now

$$\frac{Qn}{Qe} = B$$

$$Qn = BQe$$

But

$$\begin{aligned} Qe &= Qn - Qh - Q_{\text{stor}} + Q_{\text{adv}} \\ &= Qn - BQe - Q_{\text{stor}} + Q_{\text{adv}} \\ Qe + BQe &= Qn - Q_{\text{stor}} + Q_{\text{adv}} \\ (1 + B)Qe &= Qn - Q_{\text{stor}} + Q_{\text{adv}} \end{aligned}$$

Therefore

$$Q_e = \frac{Q_n - Q_{\text{stor}} + Q_{\text{adv}}}{1 + B}$$

and

$$E = \frac{Q_e}{\rho Le} = \frac{Q_n - Q_{\text{stor}} + Q_{\text{adv}}}{\rho Le(1 + B)}$$

Hence, the evaporation rate can be calculated.

If detailed data for a water body are available, the energy budget method can result in good estimates of the evaporation rate. Theoretically, the energy budget method can provide accurate estimates of E, but it requires the collection of substantial detailed energy accounting data.

EXAMPLE 4.5.6

Determine the evaporation rate, $\frac{mm}{day}$, *using the energy balance method, from a lake in northern California. The insolation flux was determined to be 650* $\frac{ly}{day}$ *and the albedo A was 8 percent for the water surface.*

The mean air temperature was 30°C and the water temperature was 21°C. The relative humidity RH was 75 percent, and longwave atmospheric radiation was estimated to be 300 $\frac{ly}{day}$ *with a reflectivity of about 3 percent.*

Back radiation was estimated to be approximately 400 $\frac{ly}{day}$. *Flow into the lake was 1-3* m^3_s *at a temperature of 20°C and outflow was 1.2* m^3_s *at 21°C. Assume that the specific heat of water is 4186* $\frac{J}{Kg\text{-}K}\left(1\ \frac{cal}{g\text{-}K}\right)$ *and that the change in stored energy is negligible. The atmospheric pressure is 102 kPa and the surface area of the lake is 500 hectares.*

Solution

$$E = \frac{Q_e}{\rho Le} = \frac{Q_n - Q_h - Q_{\text{stor}} + Q_{\text{adv}}}{\rho Le}$$

$$Qn = (1 - A)Qi + 0.97\, Q_{\text{atm}} - Q_{\text{back}}$$

$$A = 0.08; \qquad Qi = 650\,\frac{\text{ly}}{\text{day}}; \qquad Q_{atm} = 300; \qquad Q_{back} = 400\,\frac{\text{ly}}{\text{day}}$$

Therefore $\quad Qn = (1 - 0.08)650 + 0.97(300) - 400 = 489\,\dfrac{\text{ly}}{\text{day}}$

$$Q_{\text{stor}} = 0 \quad \text{(given)}$$

$$Q_{\text{adv}}\left(\frac{\text{ly}}{\text{day}}\right) = \frac{\text{Energy from inflow, }\frac{\text{cal}}{\text{day}} - \text{Energy from outflow, }\frac{\text{cal}}{\text{day}}}{\text{Surface area of lake, cm}^2}$$

Energy in flow = (density of water)(flowrate)(time − 1 day)(spec heat)(temp, K)

Since the unit of energy required here is calories, density of = water (ρ) will be in $\frac{\text{g}}{\text{cm}^3}$. Assume $\rho = 1\,\frac{\text{g}}{\text{cm}^3}$.

Although specific heat is not constant with temperature, assume constant specific heat. Note that the variation specific heat for water (at 15°C) is within 5 percent in the range of temperatures considered.

	Inflow Water	Outflow Water
Temperature:	20°C ≥ (273 + 20) = 293 K	= 21°C ≥ (273 + 21) = 294 K
Density of Water:	$1\,\frac{\text{g}}{\text{cm}^3}$	$1\,\frac{\text{g}}{\text{cm}^3}$
Flow rate:	$1.3\,\frac{\text{m}^3}{S} \geq 1.3 \times 10^6\,\frac{\text{cm}^3}{S}$	$1.2\text{ m}^3 \geq 11.2 \times 10^6\,\frac{\text{cm}^3}{S}$
Time Elapsed:	1 day ≥ 86,400S	1 day ≥ 86,400S
Lake surface area	= 500 ha	$= 5.00 \times 106\text{ m}^2$ $= 5.00 \times 1010\text{ cm}^2$

Therefore

$$Q_{\text{adv}} = \frac{(\text{Energy in} - \text{Energy out})\text{ for 1 day}}{\text{Surface area of lake, }\frac{\text{ly}}{\text{day}}}$$

$$= \left\{\left(\rho_1\left[\frac{\text{g}}{\text{cm}^3}\right]\right)\left(Q_1\left[\frac{\text{cm}^3}{S}\right]\right)(\Delta t_1\,[\text{s}])(T_1\,[\text{K}])\left(Sp\cdot ht_1\left[\frac{\text{cal}}{\text{g}}\right]\right)\right.$$

$$\left. - \left(\rho_2\left[\frac{\text{g}}{\text{cm}^3}\right]\right)\left(Q_2\left[\frac{\text{cm}^3}{S}\right]\right)(\Delta t_2\,[\text{s}])(T_2\,[\text{K}])\left(Sp\cdot ht_2\left[\frac{\text{cal}}{\text{g}}\right]\right)\right\}$$

But $\qquad p_1 = p_2, \qquad \Delta t_1 = \Delta t_2, \qquad Sp \cdot ht_1 = Sp \cdot ht_2$

Therefore
$$Q_{\text{adv}} = (\rho)(\Delta t)(Sp \cdot ht)(Q_1T_1 - Q_2T_2)$$
$$= \frac{(1)(86{,}400)(1)}{5.00 \times 10^{10}}\{(1.3 \times 10^6)(293) - (1.2 \times 10^6)(294)\}$$
$$= 48.6 \, \frac{\text{ly}}{\text{day}}$$

Atmospheric pressure $\qquad P = 102 \text{ kPa} = 1020 \text{ mb}$

The Bowen ratio,
$$B = \frac{0.66P}{1000}\left(\frac{T_s - T_a}{e_s - e_a}\right)$$

$T_s = 21°\text{C} \qquad e_s[\text{T} = 21°\text{C}] = 25.03 \text{ mb} \quad \text{(by interpolation)}$

$T_a = 30°\text{C} \qquad e_a = e_s[\text{T} = 25) \times \text{RH} = 42.43(.75)$

Relative humidity RH = 75%

Therefore
$$B = \frac{0.66(1020)}{1000}\left[\frac{21 - 30}{25.03 - 31.82}\right]$$
$$= (0.6732)(1.3254) = 0.89$$

$$\rho = \text{density of water} \cong 1 \, \frac{\text{g}}{\text{cm}^3}$$

$$Le[@\, T = 21°\text{C}] = 597.3 - 0.57T$$

$$Le = 597.3 - 0.57(21) = 585.3 \, \frac{\text{cal}}{\text{g}}$$

$$E\left(\frac{\text{mm}}{\text{day}}\right) = \frac{Q_n - Q_{\text{stor}} + Q_{\text{adv}}}{\rho Le(1 + B)}$$
$$= \frac{489 - 0 + 48.6}{(1)(585.3)(1 + 0.89)}$$

$$E = 0.486 \, \frac{\text{cm}}{\text{day}} = 4.9 \, \frac{\text{mm}}{\text{day}} \Rightarrow 5 \, \frac{\text{mm}}{\text{day}}$$

The use of the energy budget method can be a long tedious process. It requires detailed energy assessments for the watershed, the calculation of the Bowen ratio (which can be problematic when it is around -1) and a good estimate of advected heat.

4.5.6.3 Mass Transfer Method. The mass transfer method is based on the concept of the movement of water vapor from the evaporating surface to the atmosphere due to wind movement and differences in vapor pressure between the water surface and the surrounding environment. It recognizes the fact that evaporation increases with increase in wind speed and that evaporation rate increases with the dryness of the air.

A relationship that captures these factors is referred to as Dalton's law:

$$\text{Evaporation rate } E = (e_s - e_a)f(u)$$

where u = Wind speed
e_s = Saturation vapor pressure at the temperature of the water surface
e_a = Vapor pressure at some level above the water surface; usually at 2 m
$f(u)$ = Function of the wind speed. For any given system, the function, $f(u)$ is developed to account for the effect of wind speed on evaporation rate.

In other words, the function $f(u)$ is calibrated for the local environment—a given lake or reservoir.

The empirical equations developed are called *mass transfer equations*. Here are a few examples:

1. For Lake Hefner studies

$$f(u) = 0.012u_2$$

where u_2 is the wind velocity 2 m above the water surface, m/s.

2. Meyer equation

$$E\left[\frac{\text{in.}}{\text{day}}\right] = C(e_s - e_a)\left(1 + \frac{W}{10}\right)$$

where W = Wind velocity, mph, measured about 25 ft above water surface
C = Empirical coefficient
C = 0.36 for lakes
C = 0.50 for wet soil surfaces and shallow depths of water.

3. Dunne equation:

$$E = \frac{(0.013 + 0.00016u_2)\, e_a}{100}(100 - \text{RH}), \frac{\text{cm}}{\text{day}}$$

where u_2 = wind speed, $\frac{\text{km}}{\text{day}}$, at 2 m above surface
e_a = vapor pressure of air (at air temperature) in millibars
RH = relative humidity, %

Note that in implementing these methods e_s can be calculated by the following relationship when the temperature (°C) known.

$$e_s = 2.7489 \times 10^8 e^{\frac{-4278.6}{T_s + 242.79}}$$

where T_s = Temperature (°C) of water at the surface

$$e_a = (e_s \text{ @ } T_a) \times \frac{\text{Relative humidity}}{100}, (\%)$$

where T_a = Temperature (°C) of the air

4.5.6.4 Combination Methods. Two combination methods used to determine evaporation are an analytical method proposed by Penman and a graphical combination method that uses a chart.

Penman combined the energy budget method and the mass transfer method, resulting in the following equation:

$$E = \frac{\alpha En + Ea}{\alpha + 1}, \frac{\text{mm}}{\text{day}}$$

where E = Evaporation rate
α = Constant that depends on temperature given in table below called Penman's ratio
En = Evaporation due to net radiation flux absorbed by water, $\frac{\text{mm}}{\text{day}}$
Ea = Mass transfer evaporation rate, $\frac{\text{mm}}{\text{day}}$, due to dry power $(e_s - e_a)$ and wind movement $f(u)$.

The Penman equation has the advantage that the temperature of the water or soil surface does not need to be known. Because the temperature can be difficult to obtain at times it can be a significant advantage.

Values of Penman's ratio α as a function of air temperature

Air temperature T_α (°C)	$\alpha = \frac{\Delta}{\gamma}$
0	0.68
5	0.93
10	1.25
15	1.66
20	2.19
25	2.86
30	3.69
35	4.73
40	6.00

$\Delta = (0.00815T_{air} + 0.8912)^7$
For $T_{air} > -25$°C
T_{air} = Air temperature

γ = Psychometric constant $= \frac{0.66P}{1000} \cong 0.66$

P = Pressure, millibars.

EXAMPLE 4.5.7

Determine the evaporation rate E, $\frac{mm}{day}$*, from a lake in Minnesota when the air temperature is 32°C and the relative humidity is 30 percent. Net radiation flux absorbed by the body of water Qn* $= 400 \frac{ly}{day}$*. The barometric pressure is 1000 mb and the wind speed in*

the northeasterly direction is 20 miles per hour. Use the Meyer equation for Minnesota lakes to calculate the mass transfer component:

$$E\left[\frac{\text{in.}}{\text{day}}\right] = 0.0106(1 + 0.1u)(e_s - e_a)$$

where u = Wind speed in mph
$e_s - e_a$ = Saturation vapor pressure and vapor pressure, respectively, mb.

Solution 4.5.7

Radiation component of evaporation:

$$E_n = \frac{Q_n}{\rho Le}, \frac{\text{cm}}{\text{day}}$$

$$\rho = 1\,\frac{\text{g}}{\text{cm}^3}$$

$$\begin{aligned} Le &= \text{Latent heat of vaporization} \\ &= 597.3 - 0.57\text{T} = 597.3 - 0.57(32) = 579.06 \end{aligned}$$

Therefore
$$En = \frac{400\,\dfrac{\text{cal}}{\text{cm}^2 - \text{day}}}{\left(1\,\dfrac{\text{g}}{\text{cm}^3}\right)\left(579.06\,\dfrac{\text{cal}}{\text{g}}\right)} = 0.691\,\frac{\text{cm}}{\text{day}}$$

$$\alpha \text{ at } 32°\text{C} = 3.69 + (4.73 - 3.69)\frac{(32 - 30)}{(35 - 30)}$$

$$\alpha = 4.11$$

$$E_a\left[\frac{\text{in.}}{\text{day}}\right] = 0.0106[1 + 0.1(20)](e_s - e_a)$$

$$e_s \text{ (at } 32°\text{C)} = 2.7489 \times 10^8 e^{\frac{-4278.6}{32 + 242.79}} = 47.53 \text{ mb}$$

$$\begin{aligned} e_a(\text{Relative humidity}) \times e_s &= 0.30(47.53) \\ &= 014.26 \text{ mb} \end{aligned}$$

Therefore

$$E_a\left[\frac{\text{in.}}{\text{day}}\right] = 0.0106\,[1 + 0.1\,(20)](47.53 - 14.26) = 1.058\,\frac{\text{in.}}{\text{day}}$$

$$E_a\left[\frac{\text{cm}}{\text{day}}\right] = 2.54\,\frac{\text{cm}}{\text{in.}} \times 1.058\,\frac{\text{in.}}{\text{day}} = 2.69\,\frac{\text{cm}}{\text{day}}$$

$$E\left[\frac{\text{cm}}{\text{day}}\right] = \frac{\alpha En + Ea}{\alpha + 1} = 4.11(0.691) + \frac{2.69}{(4.11 + 1)} = 1.09\,\frac{\text{cm}}{\text{day}} = 10.9\,\frac{\text{mm}}{\text{day}}$$

The second method is a graphical implementation of a combination method based much on the Penman method.

Evaporation rate as a function of

- Solar radiation cal\cm^2 or ly
- Air temperature
- Dew point
- Wind movement.

Given:

1. Mean daily temp (°F)
2. Daily solar radiation absorbed, ly\days
3. Dew point temperature (°F) or relative humidity
4. Wind speed.

Procedure: Using (1) and (2), the first point is established. Using (1) and (3), the second point is established. Projecting second point down and using (4), a third point is established.

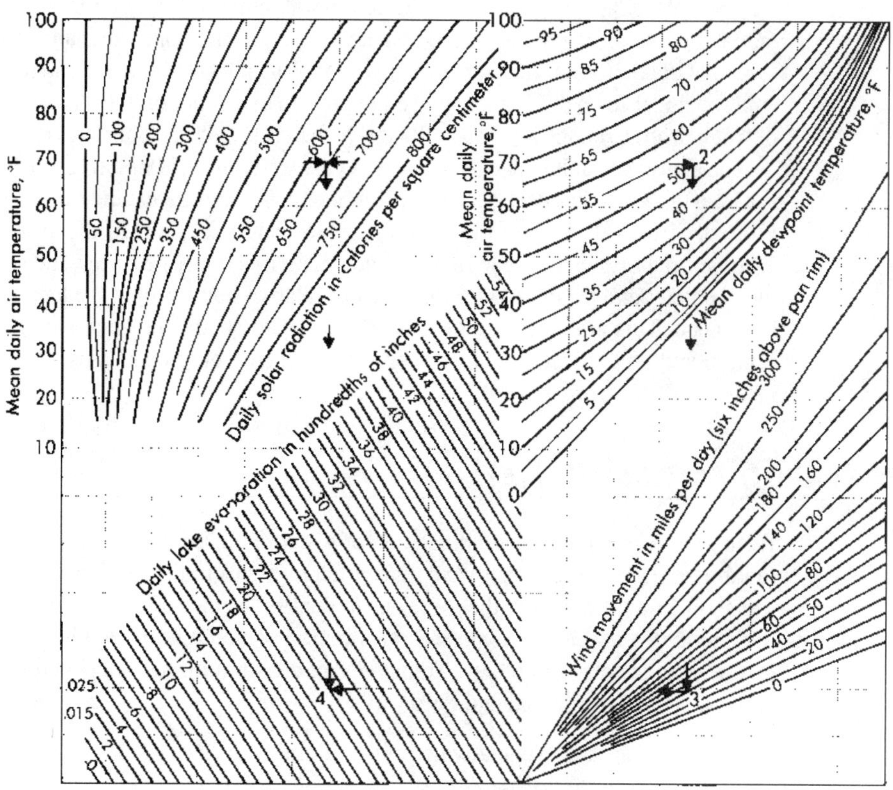

FIGURE 4.5.6 Graphical combination method based on Penman's method.

Projecting the first point vertically downward and the third point horizontally to the left; the intersection of the two lines provides the rate of evaporation in hundredths of an inch per day.

4.5.6.5 Pan Evaporation Method. In the pan evaporation method, a standard pan is filled to predetermined level with water and the evaporation (drop in level) over a given period is measured. The evaporation from the pan is called the pan evaporation. The evaporate rate to be estimated is for a body of water. The actual (lake) evaporation is related to the pan evaporation *ly* a factor called the pan coefficient;

$$E \text{ (lake)} = K_p E_p$$

where K_p = Pan coefficient
E_p = Evaporation from the pan

Evaporation pans in use include:

1. The United States National Weather Service (NWS) Class A pan
2. The Colorado sunken pan.

The NWS Class A pan is 4 ft (122 cm) in diameter and 10-in. (25.4 cm) deep.The pan, made of galvanized iron, is placed on a level open-frame platform made of wood, designed such that the bottom of the pan is about 12 in. (30.5 cm) from the ground to ensure good circulation under the pan. Water is placed in the pan to a depth of 8 in. (20 cm). When the level of water drops to about 7 in. it is refilled. The pan is monitored daily and refilled as necessary.

The Colorado sunken pan, also made of galvanized iron, is 92 cm (about 36 in) square and is 46 cm (about 18 in.) deep. The pan is set into the ground so that the rim is 5 cm (2 in.) above ground. The level of water in the pan is set to match the level of the ground.

It is a matter of practice to maintain an anemometer (to measure wind speed), rain gage, thermometer and psychrometer (to measure relative humidity) at the location of the evaporation pan. This practice ensures that relevant watershed data, necessary to correlate evaporation and other weather data, are provided for hydrologic analysis.

4.6 EVAPOTRANSPIRATION (ET)

Evapotranspiration is the combined loss of water to the atmosphere as water vapor by evaporation (as defined previously) and transpiration. Transpiration is the process by which plants transfer (draw) water from the root zone to the leaf surface through small openings (top and underside of leaf) called *stomata*. Transpiration occurs during daylight hours when photosynthesis takes place in plants. Transpiration does not directly depend on precipitation.

In short, *ET* includes evaporation

1. From water, soil, vegetative cover, and other surfaces
2. Transpiration by plants.

Evapotranspiration is influenced by the following factors:

1. Vegetation type
2. Depth of the root zone/availability of moisture
3. Rate at which plants have moisture available
4. Same factors that determine evaporation
5. Density and the extent of vegetation in the watershed.

The *field capacity* is the soil moisture content above which water is drained by gravity through the soil. Therefore, when the soil moisture content is greater than the field capacity, water percolates (seeps) downward or, otherwise, through the porous media.

The *wilting point* is the soil moisture content (level) below which plants cannot draw (extract) moisture from soil. In this situation, the pressure exerted by the plants to draw moisture is less than the soil/moisture binding pressure.

In 1948, Thornthwaite proposed the concept of Potential Evapotranspiration (*PET*). *PET* is the maximum possible *ET* rate that is limited only by weather or climatic factors. It is thought that *PET* is approximately equal to the evaporation rate from a large free-surface lake. It is possible to have conditions such that *ET* would be greater than *PET,* however, it is seldom the case. *PET*, however, provides the engineer with a target value that is used as a design parameter. Clearly, *PET* could be used to calculate the optimum crop water requirement for a vegetated watershed—typically for agricultural watersheds.

Finally, it must be noted that when soil moisture is lowered below the field capacity, actual *ET* generally falls below *PET*.

PET thus can be defined as the water loss due to evaporation (by various mechanisms) that will occur if at no time there is a deficiency of water in the soil for use by plants.

Another closely related parameter to *PET* is the concept of *consumptive water requirement* CWR—the water use rate required by a plant or crop to meet its evapotranspiration needs without any deficiencies because of lack of water. To estimate crop water requirements, a *consumptive use factor, f,* mm\day or cm\day, is multiplied by the *crop coefficient,* K_c. The crop coefficient depends on the time of year and the geographic location.

Thus,

$$ET_o = K_c f \quad \text{or} \quad q(f)$$

$$ET_c = K_c ET_o$$

where ET_c = Consumptive water requirement for a crop
ET_o = Reference crop *ET*
$q(f)$ = Some function of *f*.

4.6.1 Models for Estimating ET_o or ET

The models that have been proposed to estimate *ET* are generally empirical models and can be grouped into the following categories:

1. Temperature models
2. Radiation models

3. Combination models
4. Pan evaporation models.

(Please see previous discussion. The only difference here is that the pan coefficient will be related to the crop or plant in consideration.)

4.6.1.1 Temperature Models

1. Blaney-Criddle Model: Estimates the consumptive use factor f by a model that takes into account mean daily temperature and normalized daytime hours over a period of one month. The mathematical model is:

$$f\left[\frac{\text{mm}}{\text{day}}\right] = (0.46t + 8.13)P$$

where f = Daily consumptive use factor
t = Mean daily temperature for the month of interest (°C)
P = Mean daily daytime hours for given month\total daytime hours in the year, %.

The consumptive water required for a crop is the

$$ET\left[\frac{\text{mm}}{\text{day}}\right] = K_c f$$

where K_c = Consumptive use factor.

2. Doorenbos and Pruitt proposed that the Blaney-Criddle Model be modified to provide more accurate estimates of the reference crop evapotranspiration.
The mathematical model proposed by them is:

$$ET_o\left[\frac{\text{mm}}{\text{day}}\right] = a + bf$$

where a and b are empirical constants that depend on the minimum relative humidity RH in percent, the ratio of actual insolation time to the maximum possible insolation time $n\backslash N$, and the prevailing wind speed over a given period of time.

Table 4.6.1.1 provides values of a and b for nine categories of condition s for the Doorenbos-Pruitt adjustment of the Blaney-Criddle model.

EXAMPLE 4.6.1

Calculate the daily consumptive use factor $\left(\frac{mm}{day}\right)$ *for a location in North America at a latitude of 40°N late in the summer (August) if the mean daily temperature is 20°C. Use the Doorenbos-Pruitt correction to the Blaney-Criddle method to find the reference crop ET if a high actual insolation time, minimum relative humidity of 15 percent and the mean wind speed (2 m above the ground surface) is 3* $\frac{m}{s}$.

TABLE 4.6.1.1 Doorenbos-Pruitt Correction Factors

		Low Minimum Relative Humidity (< 20%)		Medium Relative Humidity ($20 \le RH \le 50$)		High Relative Humidity ($RH > 50\%$)	
Low Insolation Time, %, $\frac{n}{N} < 6 = 0$		a	b	a	b	a	b
	L	−2.00	1.15	−2.00	1.05	−1.45	0.80
	M	−1.80	1.28	−1.85	1.15	−1.55	0.88
	S	−1.60	1.40	−1.70	1.25	−1.65	0.98
Medium Insolation Time, %, $60 \le \frac{n}{N} \le 80$		a	b	a	b	a	b
	L	−2.30	1.35	−2.20	1.20	−1.80	0.97
	M	−2.05	1.55	−2.15	1.38	−1.75	1.06
	S	−1.80	1.73	−2.10	1.52	−1.70	1.16
High Insolation Time, %, $\frac{n}{N} > 80$		a	b	a	b	a	b
	L	−2.60	1.55	−2.40	1.37	−2.15	1.14
	M	−2.30	1.82	−2.50	1.61	−1.95	1.22
	S	−2.00	1.55	−2.55	1.82	−1.70	1.31

Table Notes:

1. $\frac{n}{N}$ = Percent actual insolation time = Actual insolation time) 100\Maximum possible insolation time

2. Wind Speed Categories

Wind Speed Categories	Wind Speed $\frac{n}{N}$ at 2 = m above ground
Light wind (L)	0–2
Moderate wind (M)	2–5
Strong wind (S)	>5

Solution

$$f\left[\frac{\text{mm}}{\text{day}}\right] = (0.46t + 8.13)P$$

$$P = 0.31, \qquad t = 20°\text{C}$$

Therefore $$f = [0.46(20) + 8.13](0.31) = 5.4\ \frac{\text{mm}}{\text{day}}$$

But $$ET_o = a + bf$$

For $$\text{RH} = 15\% \Rightarrow \text{RH} < 20\%$$

Therefore $$ET_o\left[\frac{\text{mm}}{\text{day}}\right] = 2.30 + 1.82(5.4)$$

$$= 7.5\ \frac{\text{mm}}{\text{day}}$$

3. Thornthwaite model. In the Thornthwaite model, the assumption is that evapotranspiration *ET* is related to temperature by the following type of relationship:

$$ET = aT^c$$

where a and c are constant values.

In the model, a heat index I is defined for a given month as:

$$I_n = \left(\frac{T_n}{5}\right)^{1.514} \quad \text{for a particular month } n$$

For $$n = 1, 2, \ldots, 12$$

(Note: Excluding negative mean temperatures)

where T is the mean monthly temperature, °C

The heat index I reflects Thornthwaite's classification of climates
The annual temperature efficiency index J is defined as

$$J = \sum_{i=1}^{n} I_i$$

Thus *PET* is given by

$$PET(0)\left[\frac{\text{cm}}{\text{month}}\right] = 1.62\left(\frac{10T_n}{J}\right)^{c} \quad \text{for month } n$$

where PET(0) is the PET at 0° latitude.

The number 1.62 is a value that results if each month is assumed to be 30 days long, and each day has exactly 12 hours of sunshine. This situation would occur theoretically at the equator (0° latitude).

$$C = 6.75 \times 10^{-7}J^3 - 7.71 \times 10^{-5}J^2 + 0.01792J + 0.49239$$

To adjust *PET* for locations with latitude other than 0°, a factor K is used, thus

$$PET = KPET(0)$$

The table for K values is given below in Table 4.6

EXAMPLE 4.6.2
The mean monthly temperatures for a location (40°N latitude) are given below. Determine the adjusted PET for the month of September.

TABLE 4.6.1.2 Constant K in Thornthwaite Method

Latitude	Jan	Feb	Mar	Apr	May	Jun	Jul	Aug	Sep	Oct	Nov	Dec
60°N	0.54	0.67	0.97	1.19	1.33	1.56	1.55	1.33	1.07	0.84	0.58	0.48
50°N	0.71	0.84	0.98	1.14	1.28	1.36	1.33	1.21	1.06	0.90	0.76	0.68
40°N	0.80	0.89	0.99	1.10	1.20	1.25	1.23	1.15	1.04	0.93	0.83	0.78
30°N	0.87	0.93	1.00	1.07	1.14	1.17	1.16	1.11	1.03	0.96	0.89	0.85
20°N	0.92	0.96	1.00	1.05	1.09	1.11	1.10	1.07	1.02	0.98	0.93	0.91
10°N	0.97	0.98	1.00	1.03	1.05	1.06	1.05	1.04	1.02	0.99	0.97	0.96
0°	1.00	1.00	1.00	1.00	1.00	1.00	1.00	1.00	1.00	1.00	1.00	1.00
10°S	1.05	1.04	1.02	0.99	0.97	0.96	0.97	0.98	1.00	1.03	1.05	1.06
20°S	1.10	1.07	1.02	0.98	0.93	0.91	0.92	0.96	1.00	1.05	1.09	1.11
30°S	1.16	1.11	1.03	0.96	0.89	0.85	0.87	0.93	1.00	1.07	1.14	1.17
40°S	1.23	1.15	1.04	0.93	0.83	0.78	0.80	0.89	0.99	1.10	1.20	1.25
50°S	1.33	1.19	1.05	0.89	0.75	0.68	0.70	0.82	0.97	1.13	1.27	1.36

Month	Jan	Feb	Mar	Apr	May	Jun	Jul	Aug	Sep	Oct	Nov	Dec	
Mean Temperature (°C)	−10	−15	−2	8	12	22	24	23	18	8	1	−7	Sum
Heat Index	—	—	—	2.0	3.8	9.4	10.7	10.1	7.0	2.0	0.1	0	45.1

Solution

The heat index for each month has been added to the table above. $I_i = (T_i \backslash 5)^{1.514}$

The annual temperature efficiency index $J = \sum I_i = 45.1$

For the month of September, $T = 18°C$

Therefore $$PET(0) = 1.62\left(\frac{10T_{sep}}{J}\right)^c$$

$$= 1.62\left(\frac{10(18)}{45.1}\right)^c = 1.62\left(\frac{180}{45.1}\right)^{1.206} = 8.60\,\frac{\text{cm}}{\text{month}}$$

$$c = 6.75 \times 10^{-7}(45.1)^3 - 7.71 \times 10^{-5}(45.1)^2 + 0.01792(45.1) + 0.49239$$
$$= 1.206$$

$$PET\text{ (at 40°N latitude)} = KPET(0) = 1.04(8.60) = 8.94\,\frac{\text{cm}}{\text{month}}$$

$K = 1.04$ in September at 40°N latitude in Table 4.5

4.6.1.2 Radiation Models. Priestley and Taylor proposed a model based on the assumption that *PET* can be estimated as the adjusted value of the radiation component of the combination model for estimating surface water evaporation proposed by Penman (see the previous section). The mass transfer component is assumed to be zero ($E_a = 0$)

The Priestley and Taylor model is:

$$PET\left[\frac{\text{cm}}{\text{day}}\right] = \frac{1.26\alpha Q_n}{\rho Le(\alpha + 1)}\left(\frac{180}{45.1}\right)$$

where Q_n = Absorbed insolation flux, $\frac{\text{ly}}{\text{day}}$

$\alpha = \frac{\Delta}{\gamma}$ = A constant from Table 4.2

ρ = Density of water at mean temperature $\left(\text{usually assumed} = 1\,\frac{\text{g}}{\text{cm}^3}\right), \frac{\text{g}}{\text{cm}^3}$

Le = Latent heat of vaporization for water, $\frac{\text{cal}}{\text{g}}$.

4.6.1.3 Combination Models

1. Jensen-Haise Method

Jensen-Haise analyzed evaporation data in about 3000 cases using a soil sampling technique over a period of 35 years. Their work led them to propose the following combination model for estimating evapotranspiration.

$$ET\left[\frac{\text{mm}}{\text{day}}\right] = \frac{10C_T(T - T_x)Q_n}{\rho Le}$$

where Q_n = Net insolation flux, $\frac{\text{ly}}{\text{day}}$

T = Mean air temperature, °C

T_x = Reference temperature index calculated as a function of elevation above sea level and a vapor pressure differential (see below), °C

C_T = Temperature constant; approximately 0.025 (when data to calculate C_T is not available, 0.025 may be assumed)

$$C_T = \frac{1}{C_1 + C_2 C_H} \qquad \text{and} \qquad C_H = \frac{50}{e_2 - e_1}, \text{ mb}$$

where $C_2 = 7.6°\text{C}$

$$C_1 = 38 - \frac{2\ [°\text{C}] \times \text{Elevation [m]}}{305}$$

e_1 = Saturation vapor pressure at mean minimum for the warmest month of the year, mb

e_2 = Saturation vapor pressure at the mean maximum temperature for the warmest month of the year, mb

$$T_x = -\ 2.5 - 0.14\ e_2 - e_1 - \frac{\text{elevation [m]}}{550}$$

ρ = Density of water mean air temperature, $\frac{\text{g}}{\text{cm}^3}$

Le = Latent heat of vaporization of water at the mean air temperature, $\frac{\text{cal}}{\text{g}}$

EXAMPLE 4.6.3

At a certain location in the northern hemisphere at latitude 30°, the mean air temperature is 25°C and the average incident insolation is 810 $\frac{ly}{day}$ with an albedo of 8 percent. The mean maximum and minimum temperatures for the warmest month are 30°C and 20°C, respectively. The elevation above mean sea level is 200 m.

Determine the potential evapotranspiration, $\frac{mm}{day}$, using the Jensen-Haise method.

Solution

First, determine the density (ρ) and latent heat of vaporization Le at the mean temperature (15°C) from Reference Table

$$\rho = 0.99705\ \frac{\text{g}}{\text{cm}^3}$$

$$Le = 583.2\ \frac{\text{cal}}{\text{g}}$$

Now

$$C_T = \frac{1}{C_1 + C_2 C_H}; \qquad C_1 = 38 - 2 \times \frac{\text{Elev, m}}{305}$$

$$C_2 = 7.6$$

Therefore
$$C_T = \frac{1}{36.69 + (7.6)(2.62)} = 38 - 2 \times \frac{200}{305}$$
$$CT = 0.01766 \qquad C_1 = 36.69$$
$$e_2 \text{ @ } T = 30°\text{C} = 42.43 \text{ mb}$$
$$e_1 \text{ @ } T = 20°\text{C} = 23.37 \text{ mb}$$
$$C_H = \frac{50}{e_2 - e_1} = \frac{50}{42.43 - 23.37} = \frac{50}{19.06} = 2.62$$
$$T_x = -2.5 - 0.14\, e_2 - e_1$$
$$= -2.5 - 0.14(42.43 - 23.37)$$
$$T_x = -5.17$$

But
$$PE\left[\frac{\text{mm}}{\text{day}}\right] = \frac{10C_T(T - T_x)Q_n}{\rho Le}$$
$$\text{Incident insolation} = 810\,\frac{\text{ly}}{\text{day}}$$
$$\text{Absorbed insolation} = (1 - \text{albedo})(\text{incident insolation})$$

Therefore
$$Q_n = (1 - 0.08)(810) = 745.2\,\frac{\text{ly}}{\text{day}}$$

Substituting values:
$$PET = \frac{10(0.01766)[25 - (-5.17)]745.2}{(0.99705)(583.2)}$$
$$= 6.8\,\frac{\text{mm}}{\text{day}}$$

2. Turc Method. Turc analyzed rainfall, evaporation, and other data from 254 watersheds around the world. From the analyses, Turc proposed a model that relates evaporation to rainfall and temperature.

   Turc proposed the following:
$$\text{Annual evaporation } E\left[\frac{\text{mm}}{\text{day}}\right] = \frac{P}{\left[0.9 + \left(\frac{P}{I_T}\right)^2\right]^{\frac{1}{2}}}$$

where P = Annual precipitation amount, mm
$I_T = 300 + 25T + 0.05T^3$
T = Mean air temperature, °C.

In addition, Turc proposed regional models for estimating *ET* under humid or arid conditions.

For Western Europe:

1. When relative humidity $RH \geq 50\%$

$$ET\left[\frac{\text{mm}}{\text{day}}\right] = \frac{0.013T(Q_n + 50)}{T + 15}$$

2. When relative humidity $RH < 50\%$:

$$ET\left[\frac{\text{mm}}{\text{day}}\right] = \frac{0.013T(Q_n + 50)}{T + 15}\left(1 + \frac{50 - R_H}{70}\right)$$

where Q_n = Net insolation flux, $\frac{\text{ly}}{\text{day}}$
T = Mean air temperature, °C

4.7 PEAK RUNOFF MODELS

The models used to simulate the rainfall runoff relationship are presented in this and the following sections. In general, the two classes of models are peak runoff models and continuous runoff models.

Peak runoff models provide the peak flow rate from a watershed; continuous runoff models provide the flow rate and time relationship. This flow-time relationship or curve is called the runoff or storm hydrograph.

Two peak runoff models, which provide the peak runoff rate from the watershed, are:

1. Rational method
2. NRCS TR-55 method.

4.7.1 Rational Method

The rational method, called the Lloyd-Davies method in England, has been used widely to calculate the peak runoff from a small watershed. The rational method provides the peak runoff rate from a catchment given the runoff coefficient C, the time of concentration tc, the area of the catchment or watershed, and information to calculate the input or design storm or rainfall event.

The following assumptions are made when applying the rational method:

1. The catchment or watershed is small.
2. The catchment is concentrated—that is, the rainfall duration is longer than, or at least equal to, the time of concentration of the watershed.
3. The rainfall event is spatially and temporally constant over the duration and area of the watershed, respectively.
4. The runoff coefficient is a catch-all coefficient that incorporates all losses (hydrologic abstractions) and diffusive properties of the catchment.

Because a constant rainfall intensity and a constant area coverage of the rainfall are assumed, the catchment should be small for the proper application of the rational method. In general, the time of concentration is assumed to be less than one hour and a maximum watershed size of 2.5 km^2 (250 hectares or 640 acres) is used as a guide. Other sources have noted a limit of 20 minutes for tc and an area upper limit of 80 hectares (200 acres). Still, these are suggestions of even more stringent limits on these parameters.

The rational method determines the peak runoff rate:

$$Q_p\left(\frac{\text{ft}^3}{\text{s}}\right) = CIA$$

where C = Runoff coefficient
I = Rainfall intensity, $\frac{\text{in.}}{\text{hr}}$
A = Area of watershed, acres

The rational method is a simple and straightforward method for determining the maximum discharge from a watershed. It must be applied with caution, however, ensuring that the underlying assumptions implied in its development are not violated or entirely disregarded. An obvious weakness of the rational method is that it does not consider the antecedent moisture condition of a watershed.

Implementation Steps

1. Determine the area of the watershed
2. Determine the length and slope for the catchment and calculate the time of concentration (t_c). Because an equilibrium condition is assumed, the watershed must be concentrated, which means that the duration of the design storm is at least as long as the time of concentration (t_c). Thus, rainfall duration = t_c. Given t_c, use the appropriate intensity-duration frequency curve (IDF curve) to determine the design rainfall intensity for the appropriate return period (which indicates the level of protection or acceptable risk level).
3. Use appropriate tables to determine the runoff coefficient, C. If the watershed is used for several classes of land-use applications, a weighted C value should be computed first. Otherwise, a modified form of the rational method, which is

$$Q_{\text{peak}}\left(\frac{\text{ft}^3}{\text{s}}\right) = I\left(\sum_{i=1}^{n} C_i A_i\right)$$

where C_i = Runoff coefficient for land-use application i
A_i = Corresponding area in acres.

Or

$$C = \frac{\sum_{i=1}^{n} C_i A_i}{\sum_{i=1}^{n} A_i}$$

Any of the following formulas may be used with the appropriate units:

a.
$$Q_p\left(\frac{\text{ft}^3}{\text{s}}\right) = 1.008CIA$$

where 1.008 usually is taken to be 1.0
C = Runoff coefficient
I = Rainfall intensity, $\frac{\text{in.}}{\text{hr}}$
A = Area of watershed, acres

b.
$$Q_p\left(\frac{\text{ft}^3}{\text{s}}\right) = (2.314)(10^{-5}CIA)$$

where I = Rainfall intensity, $\frac{\text{in.}}{\text{hr}}$
A = Area of watershed, ft^2

c.
$$Q_p\left(\frac{\text{m}^3}{\text{s}}\right) = 0.278CIA$$

where I = Rainfall intensity, $\frac{\text{mm}}{\text{hr}}$
A = Area of watershed, km^2

d.
$$Q_p\left(\frac{\text{m}^3}{\text{s}}\right) = (2.78)(10^{-3}CIA)$$

EXAMPLE 4.7.1.1

Determine the peak discharge from a watershed in Charlotte, NC, for a 10-year design storm for the conditions shown in the watershed shown in Figure 4.7.1.1a. The length of the watershed, from the watershed divide (elevation 660 ft) to the outlet (elevation 656 ft) along the main channel in the watershed is 600 ft and the area is 150 acres. The minimum time of concentration (by ordinance) is 15 minutes.

The land-use applications in the watershed are:

Woodland	*50%*
Paved areas (asphalt)	*10%*
Lawn/open spaces	*40%*

Solution

For Charlotte, NC, the IDF curve for a 10-year storm is:

$$i\left(\frac{\text{mm}}{\text{hr}}\right) = \frac{4851}{t + 22}$$

where t is the duration of the rainfall in minutes.

$$\text{The slope of the watershed} = \frac{660 - 656}{600} = 0.0067\ \frac{\text{ft}}{\text{ft}}$$

Time (hrs)	Capacity, f (mm/hr)
00.00	03.500
00.10	03.242
00.20	03.006
00.40	02.593
00.80	01.960
01.00	01.720
02.00	00.996
03.00	00.702
04.00	00.582
05.00	00.533
06.00	00.514
07.00	00.506
08.00	00.502
09.00	00.501
10.00	00.500
15.00	00.500
20.00	00.500

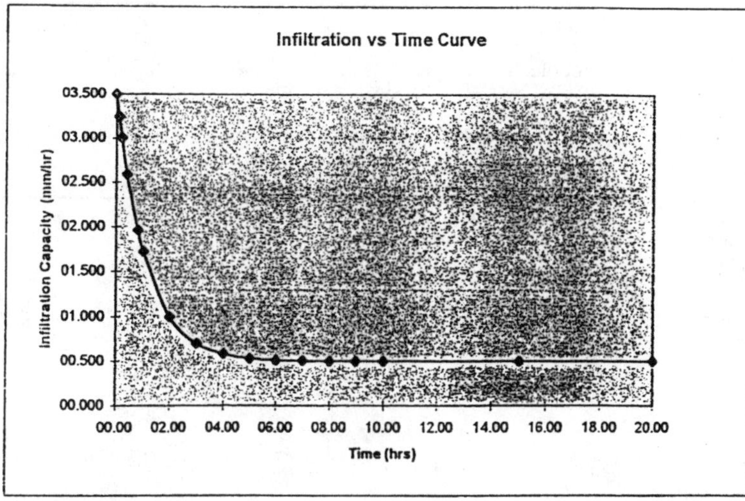

FIGURE 4.7.1.1A Watershed for Example 4.7.1.1a.

Using the Kirpich formula to calculate the time of concentration:

$$t_c = \frac{0.0078L^{0.77}}{S^{0.385}} = \frac{0.0078(600)^{0.77}}{(0.0067)^{0.385}} = 7.4 \text{ minutes}$$

Because the minimum t_c (by ordinance) is 15 minutes,

$$t_c = 15 \text{ minutes}$$

therefore

$$i\left(\frac{\text{mm}}{\text{yr}}\right) = \frac{4851}{15 + 22} = 131.1\left(\frac{\text{mm}}{\text{hr}}\right)$$

$$\text{Area, } A = 150 \text{ acres} = \left(0.4047\,\frac{\text{hectare}}{\text{acre}}\right)(150 \text{ acres}) = 60.71 \text{ hectare}$$

Computing the composite runoff coefficient:

$$\text{Weighted } C = \frac{\sum_{i=1}^{n} A_i C_i}{\sum_{i=1}^{n} A_i}\ \sum_{i=1}^{n}\left(\frac{A_i}{A}\right)C_i = \sum_{i=1}^{n} P_i A_i$$

where

$$P_i = \frac{\text{fraction of the } i\text{th land use}}{\text{application}}$$

TABLE 4.7.1.1A *C* Calculation

Land use	C_i	P_i	Remarks
Woodland	0.2	0.50	Since soil type was not specified, sandy to clay soil is assumed.
Paved area (asphalt)	0.85	0.10	*C* from Table 4.7.4
Lawn/open areas < 1% slope	0.10	0.40	*C* from Table 4.7.4

C_i was obtained from Tables 4.7.4 and 4.7.5.

Therefore, $$C = 0.5(.2) + .85(.10) + 0.10(40) = 0.225$$

Therefore, $$Q\left(\frac{\text{m}^3}{\text{s}}\right) = (2.78)(10^{-3}CIA)$$

$$= (2.78)(10^{-3})(0.225)\left(131.1\ \frac{\text{mm}}{\text{hr}}\right)(60.71\ \text{hr})$$

$$Q_{\text{peak}} = 4.98\ \frac{\text{m}^3}{\text{s}}$$

4.7.2 NRCS TR-55 Method

Based on results obtained from rainfall-runoff computations in small and mid-sized watersheds, the Natural Resources Conservation Service (NRCS) proposed the TR-55 method for determining the peak runoff from a catchment.

By the TR-55 method, peak discharge from a watershed is estimated by:

$$Q_{\text{peak}}\left(\frac{\text{m}^3}{\text{s}}\right) = quAQF$$

The TR-55 method estimates peak flow from a watershed using the concept of unit peak flow, *qu* (m^3s) per cm per km^2

where A = Watershed area km^2
Q = Runoff depth from a 24-hour storm for a given return period, cm
F = Surface storage or pond and swamp adjustment factor (see Table 4.7.2.1)

In this peak runoff model, the unit peak flow, qu depends on

1. Time of concentration
2. NRCS storm type (Type I, Type IA, Type II and Type III)
3. Ratio of the abstraction at the onset of the storm (*Ia*) to the total rainfall during the storm (*P*).

The runoff depth, Q is computed directly from the NRCS curve number (*CN*) relationship where *P* is the total 24-hour rainfall.

The curve number method is discussed fully in Section 4.8.1. The reader, however, should confirm that:

$$\text{Potential retention, } S \text{ (mm)} = 25.4\left(\frac{1000}{Cn - 10}\right)$$

And that initial abstraction (Ia) is estimated to be 0.25.

And finally depth of runoff:

$$Q \text{ (mm)} = \frac{P - 0.2S}{P + 0.8S}$$

The empirical relationship:

$$\log q_u = C_0 + C_1 \log t_c + C_2(\log t_c)^2 - 2.366$$

(This equation is used in lieu of NRCS charts of t_c, $\frac{Ia}{P}$ and q_u)

C_0, C_1 and C_2 are provided in Table 4.7.1.1 and t_c is the time of concentration (hrs). However, $0.1 \leq t_c \leq 10$ hours.

In Table 4.7.2.2, if $\frac{Ia}{P}$ is less than 0.1, use values for $\frac{Ia}{P} = 0.10$ and if $\frac{Ia}{P} > 0.5$, use values for $\frac{Ia}{P} = 5.0$.

Pond and swamp areas exceeding 5 percent may require routing runoff through storage areas. Typically, pond and swamp areas should be distributed over the watershed.

The NRCS TR-55 method does not provide good estimates of peak flow if the watershed is not nearly homogeneous. To be considered homogeneous, adjacent CN values should be within 5 points of each other, the overall CN for the catchment should be greater than 40, the time of concentration tc should be nearly the same for all main channels, and its value should be between 6 minutes and 10 hours.

EXAMPLE 4.7.2.1

A 3.5 km² watershed has an average CN value of 8. Given a time of concentration of 2-3 hrs, determine the 25-year peak runoff from a NRCS Type IA storm if the total rainfall depth is 12 cm and the percentage of pond and swamp area is 3 percent.

TABLE 4.7.2.1 NRCS TR-55 Pond and Swamp Areas (Surface Storage) Correction Factor

Percentage of Pond and Swamp Areas	F
0.0	1.00
0.2	0.97
1.0	0.87
3.0	0.75
5.0	0.72

TABLE 4.7.2.2 Parameters for Estimating the Unit Peak Discharge, *qu*

NRCS Rainfall Type	$\frac{Ia}{P}$	C_0	C_1	C_2
I	0.10	2.30550	−0.51429	−0.11750
	0.20	2.23537	−0.50387	−0.08929
	0.25	2.18219	−0.48488	−0.06589
	0.30	2.40624	−0.45695	−0.02835
	0.35	2.00303	−0.40769	0.01983
	0.40	1.87733	−0.32274	0.05754
	0.45	1.76312	−0.15644	0.00453
	0.50	1.67889	−0.06930	0.0
IA	0.10	2.03250	−0.31583	−0.13748
	0.20	1.91978	−0.28215	−0.07020
	0.25	1.83842	−0.25543	−0.02597
	0.30	1.72657	−0.19826	0.02633
	0.50	1.63417	−0.03180	0.0
II	0.10	2.55323	−0.61512	−0.16403
	0.30	2.46532	−0.62257	−0.11657
	0.35	2.41896	−0.61594	−0.08820
	0.40	2.36409	−0.59857	−0.05621
	0.45	2.29238	−0.57005	−0.02281
	0.50	2.20282	−0.51599	−0.01259
III	0.10	2.47317	−0.51848	−0.17083
	0.30	2.39628	−0.51848	−0.17083
	0.35	2.35477	−0.49738	−0.11985
	0.40	2.30726	−0.46541	−0.11094
	0.45	2.24876	−0.41314	−0.11508
	0.50	2.17772	−0.36803	−0.09525

Solution

Potential retention $$S = 25.4\left(\frac{1000}{Cn - 10}\right)$$

$$= 25.4\left(\frac{1000}{78 - 10}\right) = 71.64 \text{ mm}$$

Total precipitation $$P = 12 \text{ cm} = 120 \text{ mm}$$

Using the NRCS curve number method:

Runoff depth, $$Q = \frac{[P - 0.2S]^2}{P + 0.8S} = \frac{[120 - 0.2(71.640)]^2}{120 + 0.8(71.64)} = 63.0 \text{ mm}$$

$$= 6.3 \text{ cm}$$

From Table 4.7.2.1, since the percentage of pond area is 3%, $F = 0.75$

But $$Ia = 0.25 = 0.2(71.64 \text{ mm}) = 14.3 \text{ mm}$$

Therefore, $$\frac{Ia}{P} = \frac{14.3}{120} = 0.12$$

(Good, since it is between 0.10 and 0.50.)

In addition, $$t_c = 2.3 \text{ hrs}$$

(Good, since $0.1 \leq t_c \leq 10$ hrs.)

From Table 4.7.2.2, determine C_0, C_1, and C_2.

For Type IA storm (by interpolation):

$C_0 = 2.29147$

$C_1 = -0.51221$

$C_2 = -0.11186$

But,

$$\begin{aligned}\log qu &= C_0 + C_1 \log t_c + C_2(\log t_c)^2 - 2.366\\ &= 2.29147 + [-0.51221(0.361728)] + [-0.11186(0.361728)2] - 2.366\\ &= -\ 0.27445\end{aligned}$$

$$qu = 10^{-0.27445} = 0.532 \frac{\text{m}^3}{\text{s}} \text{ per cm per km}^2$$

Therefore, the unit peak discharge $qu = 0.532 \left(\frac{\text{m}^3}{\text{s}}\right)$ per cm per km^2

$$\begin{aligned}Q_{\text{peak}}\left(\frac{\text{m}^3}{\text{s}}\right) &= quAQF\\ &= (0.532)(3.5 \text{ km}^2)(6.3 \text{ cm})(0.75)\\ &= 8.79 \frac{\text{m}^3}{\text{s}}\end{aligned}$$

4.8 CONTINUOUS RUNOFF MODELS

This class of rainfall-runoff models provides the storm hydrograph for a given rainfall hyetograph. In contrast to peak runoff models, which provide only the estimate of the peak discharge from the watershed, these models provide an estimate of the runoff versus time series. The four types of continuous runoff models used in engineering hydrology are

1. Time-area methods
2. Unit hydrograph techniques, including synthetic hydrographs
3. Kinematic wave models
4. Non-linear reservoir models.

Only time-area methods and unit hydrograph methods will be treated in this section.

4.8.1 Time-Area Models

The idea or concept of isochrones is central to the understanding of the time-area model. Rainfall excess or runoff is produced when rainfall intensity is higher than total hydrologic abstractions (depression storage, interception, infiltration, and evaporation). The runoff hydrograph that results at the watershed outlet essentially is due to runoff contributions, arriving at the outlet at different times, as contributions of overland flow and flow through any established channels from all locations in the watershed. An isochrone is the locus of points that have equal travel or arrival time (for runoff) at the watershed outlet.

The concept of the time-area histogram is in addition to, or as part of, the concept of isochrones. Assuming that outflow at the watershed outlet results from only the translation of rainfall excess (concentrating from the watershed) to the watershed outlet, without regard to any storage effects in the watershed, the time-area histograph concept can determine the resulting hydrograph. The rainfall hyetograph translates in time in accordance to the time-area histogram to the watershed outlet, resulting in the runoff hydrograph.

Thus if the rainfall excess R is described in the same time increments as the time-area histogram, as average rainfall (excess) intensity, then runoff at time increment Q_j is given by

$$Q_j = R_jA_1 + R_{j-1}A_2 + R_1A_j$$

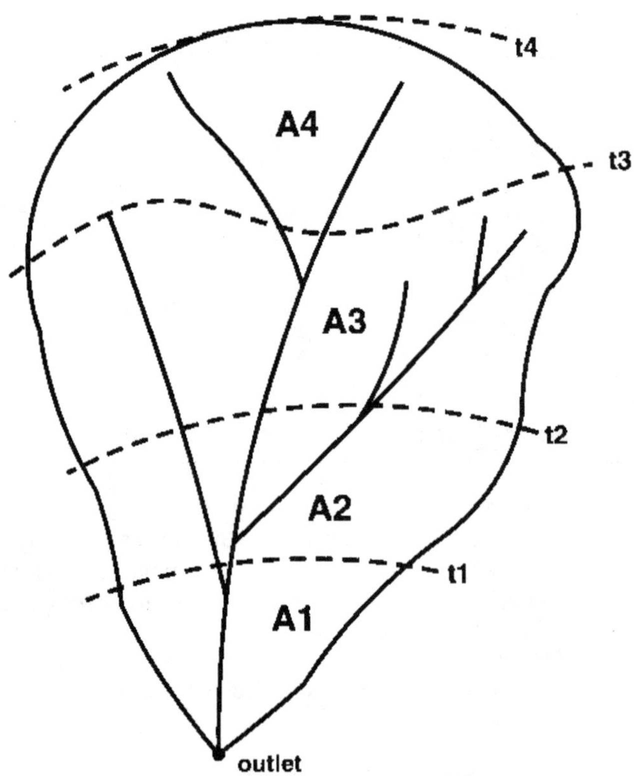

FIGURE 4.8.1.1 Watershed showing runoff isochrones.

where R_1 is the rainfall excess intensity.

If $\quad Q_j\left(\frac{\text{m}^3}{\text{s}}\right) \quad$ and $\quad R_j\left(\frac{\text{mm}}{\text{hr}}\right) \quad$ and $\quad A_j$ (hectare),

then
$$Q_j\left(\frac{\text{m}^3}{\text{s}}\right) = 0.002778\,(R_jA_1 + R_{j-1}A_2 + R_1A_j)$$

If $\quad Q_j\left(\frac{\text{ft}^3}{\text{s}}\right) \quad$ and $\quad R_j\left(\frac{\text{in.}}{\text{hr}}\right) \quad$ and $\quad A_j$ (acres)

then
$$Q_j\left(\frac{\text{ft}^3}{\text{s}}\right) = 1.008\,(R_jA_j + R_{j-1}A_2 + R_1A_j)$$

The 1.008 usually is taken to be 1.00

EXAMPLE 4.8.1.1

Use the time-area method to develop the hydrograph (tabulate your values) for a watershed during a four-hour storm. The area of the watershed is 27 acres and the ϕ*-index is*

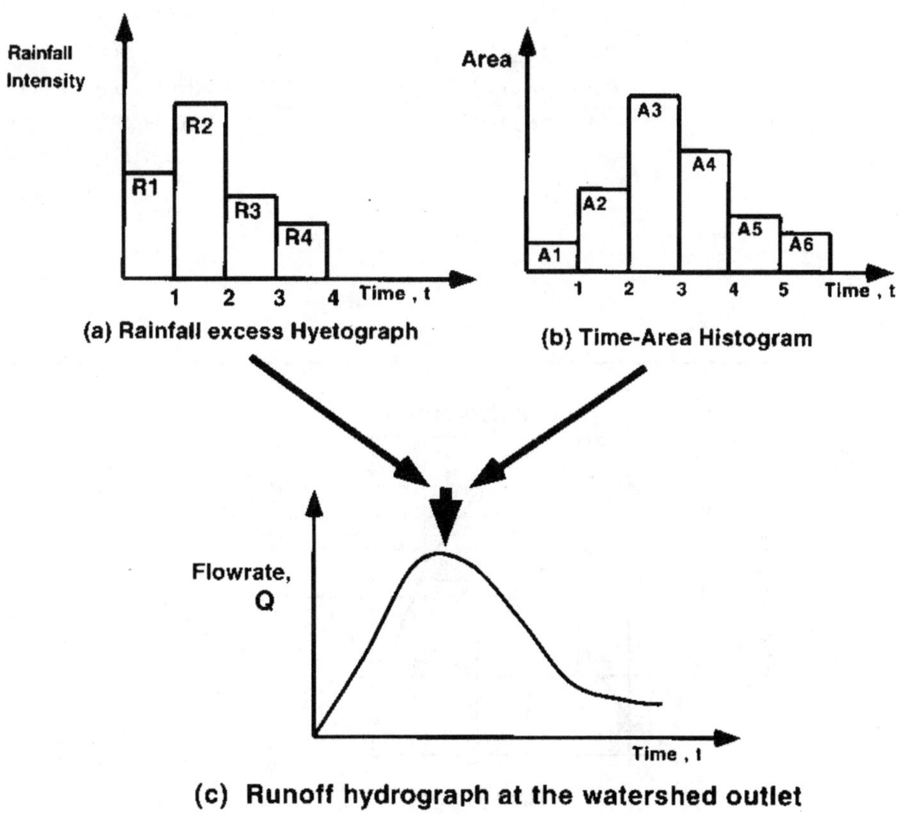

FIGURE 4.8.1.1A Time-area method elements.

0.50 in./hr. Begin by completing the rainfall excess column in the table below by taking into account the abstraction due to infiltration. Next, determine the flow rate for each period in the second table provided.

Rainfall distribution		Effective Rainfall ordinates	Time-Area Histogram	
Time (hrs)	Rainfall Rate $\left(\frac{in}{hr}\right)$	Rainfall Excess Rate $\left(\frac{in}{hr}\right)$	Time to Outlet (hrs)	Area (acres)
1	1.0	0.5	1	4
2	1.5	1.0	2	6
3	1.0	0.5	3	10
4	0.5	0.0	4	7

Time (hrs)	Detail of Q_i calculations		Flowrate (cfs)
0	0.0		0
1	R_1A_1	$= 0.5(4)$	2.0
2	$R_2A_1 + R_1A_2$	$= 1.0(4) + 0.5(6)$	7.0
3	$R_3A_1 + R_2A_2 + R_1A_3$	$= 0.5(4) + 1(6) + 5(10)$	13.0
4	$R_4A_1 + R_3A_2 + R_2A_3 + R_1A_4$	$= 0(4) + 5(6) + 1(10) + 5(7)$	16.5
5	$R_4A_2 + R_3A_3 + R_2A_4 =$	$0(6) + 5(10) + 1(7)$	12.0
6	$R_4A_3 + R_3A_4 =$	$0(10) + 5(7)$	3.5
7	$R_4A_4 =$	$0(7)$	0.0
8			
9			
10			
11			

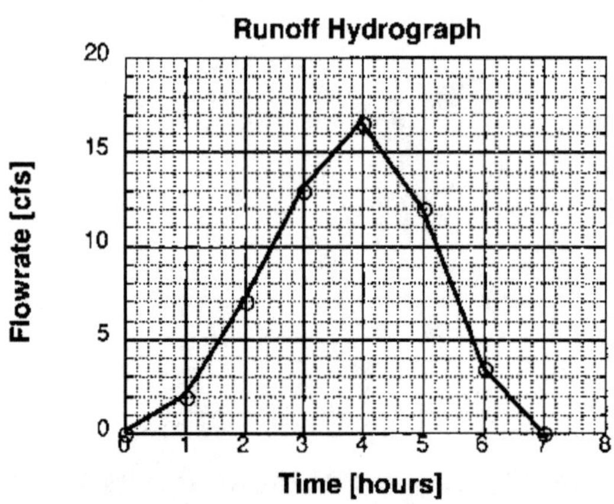

FIGURE E4.8.1.1A Runoff hydrograph using time-area method.

4.8.2 Unit Hydrograph Theory

The concept of the unit hydrograph is one of the most important concepts in relation to estimating a runoff hydrograph from a rainfall hyetograph. In fact, the time-area method, discussed in the previous section, is a special case of the unit hydrograph method.

The unit hydrograph method is based on three principles:

1. Linearity of the watershed response to rainfall excess.
2. The idea that rainfall excess timing can be logged. Response is the same for the same amount of rainfall excess; timing may be shifted (lagged) to allow for the timing of rainfall excess.
3. The concept of super position. Rainfall excess response can be superimposed (added together in time).

A hydrograph is a continuous time-discharge curve (usually unimodal) that describes the hydrologic behavior or response of a watershed to rainfall.

A unit hydrograph is defined as the discharge curve resulting from a unit depth (one inch or 1 centimeter) of rainfall excess, which is distributed uniformly over the watershed at a constant rainfall rate during a specified period of rainfall duration. The properties or characteristics of a unit hydrograph are:

1. The volume under a unit hydrograph is equal to 1 unit (1 cm or 1 in.) rainfall excess
2. If the duration of two rainfall excess events is equal, without regard to their respective rainfall intensities, they must result in the same hydrograph time base.
3. The unit hydrograph results in a linear system whereby the direct runoff for a storm of a specified duration is directly proportional to the rainfall excess amount or volume.
4. Rainfall distribution for all equal duration storms is identical in space and time.

The S-Curve Method. The S-curve method can be used to generate unit hydrographs of larger or smaller duration by taking advantage of the linear property of the unit hydrograph. In fact, the S-curve method provides a means for generating a unit hydrograph of any duration (given a particular unit hydrograph).

Procedure. Given a unit hydrograph (UH) of duration D0 and desiring to generate another unit hydrograph (UH) of duration D1.

1. Generate an S-curve by superimposing (adding) a series of UHs of duration D each lagged in time by the duration D. Lag as many UHs as needed until the upper right tail of the S-curve is virtually horizontal.
2. Shift the S-curve in time by the new duration D1
3. Subtract corresponding ordinates of the original S-curve and the shifted S-curve.
4. Multiply the result of step 3 by the ratio D0/D1. This produces the desired UH with duration D1.

EXAMPLE 4.8.2.1

Determine the three-hour unit hydrograph from the one-hour unit hydrograph given in the table using the S-curve method. Plot the original (one-hour) and the three-hour unit hydrographs, as well as the S-curves used to generate the three-hour UH.

Solution

The S-curve is generated by continuously lagging and adding 17 one-hour unit hydrographs. A timc pcriod of 0-14 hours was used for the subsequent calculations.

The necessary calculations and the graphical representation of the results have been provided in Table E4.8.2.1a and Figure E4.8.2.1.

Hydrograph Convolution: Generating a Storm Hydrograph. The unit hydrograph technique is powerful because of the simple method for generating the continuous runoff hydrograph given an input rainfall excess hyetograph and the appropriate UH.

The process of multiplying a given UH by a rainfall excess distribution to generate a storm hydrograph is called hydrograph convolution.

In short, the rainfall excess is taken period by period, multiplied with the UH and the corresponding hydrograph ordinates are added to generate the storm hydrograph. This process can be facilitated by the use of a spreadsheet.

EXAMPLE 4.8.2.2

Using the one-hour UH of Example 4.8.2.1, determine the runoff or storm hydrograph for the following rainfall excess.

Time (hr)	1	2	3	4
Rainfall Excess $\left(\frac{\text{cm}}{\text{hr}}\right)$	0.5	1.2	2.5	0.8

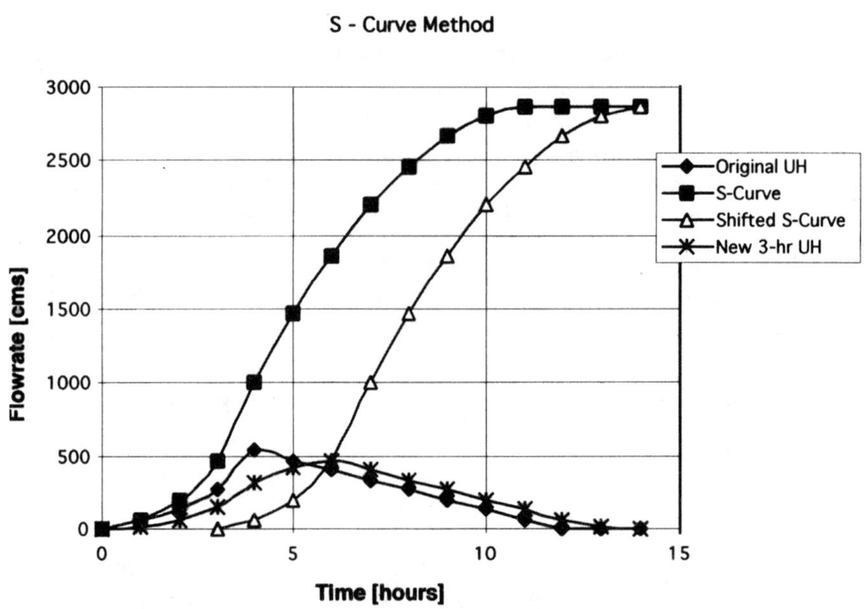

FIGURE E4.8.2.1A S-curve and unit hydrographs for Example 4.8.2.1.

TABLE E4.8.2.2A Computations for S-curve Method

(1)	(2)	(3)	(4)	(5)	(6)
Time (hrs)	1-hour UH (cms)	S-curve (cms)	Shifted S-curve (cms)	Difference (3) − (4) (cms)	(5) × $\left(\frac{1}{3}\right)$ 3-hr UH (cms)
0	0	0		0	0
1	66	66		66	22
2	134	200		200	67
3	266	466	0	466	155
4	534	1000	66	934	311
5	466	1466	200	1266	422
6	400	1866	466	1400	467
7	334	2200	1000	1200	400
8	266	2466	1466	1000	333
9	200	2666	1866	600	267
10	134	2800	2200	600	200
11	66	2866	2466	400	133
12	0	2866	2666	200	67
13	0	2866	2800	66	22
14	0	2866	2866	0	0

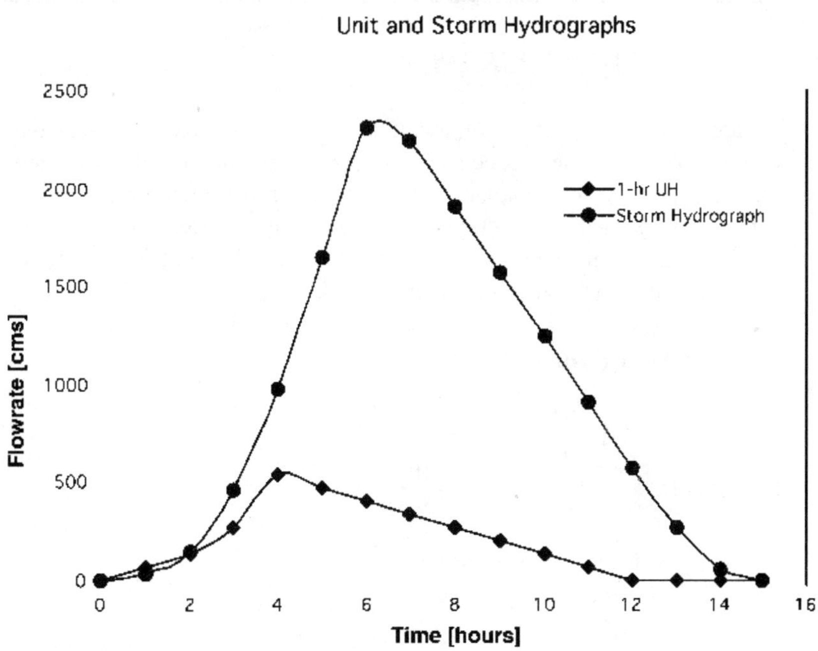

FIGURE E4.8.2.2A Unit and storm hydrographs for Example 4.8.2.2.

TABLE E4.9.2.2

(1)	(2)	(3)	(4)	(5)	(6)	
	1-hour		Rainfall Excess $\left(\frac{cm}{hr}\right)$			Storm Hydrograph
Time	UH					(3) + (4) + (5) + (6)
(hrs)	(cms)	0.5	1.2	2.5	0.8	(cms)
0	0	0				0
1	66	33	0			33
2	134	67	79.2	0		146.2
3	266	133	160.8	165	0	458.8
4	534	267	319.2	335	52.8	974
5	466	233	640.8	665	107.2	1646
6	400	200	559.2	1335	212.8	2307
7	334	167	480	1165	427.2	2239.2
8	266	133	400.8	1000	372.8	1906.6
9	200	100	319.2	835	320	1574.2
10	134	67	240	665	267.2	1239.2
11	66	33	160.8	500	212.8	906.6
12	0	0	79.2	335	160	574.2
13	0	0	0	165	107.2	272.2
14	0	0	0	0	52.8	52.8
15	0	0	0	0	0	0

4.9 SYNTHETIC UNIT HYDROGRAPHS

In order to apply the unit hydrograph technique to a particular watershed, the appropriate unit hydrograph must be developed by using appropriate data to calibrate the unit hydrograph for the watershed. Such data are available only for properly instrumented watersheds. For ungaged sites, synthetic hydrographs are the alternative process for developing an appropriate unit hydrograph. In general, synthetic unit hydrograph methods allow for the application of the technique using generalized methods.

NRCS Dimensionless Unit Hydrograph

The NRCS unit hydrograph is defined by the following relationships:

Peak runoff,
$$q_p\left(\frac{\text{ft}^3}{\text{s}}\right) = \frac{484AQ}{t_p} = \frac{726AQ}{t_c}$$

where Q = Runoff in inches (described below)
A = Area in square miles
t_p = Time to peak runoff (hours), $\frac{2}{3}t_c$
t_c = Time of concentration.

The NRCS curve number method provides a relationship between the time lag (t_L) and the time of concentration, t_c.

Therefore, $$t_L = 0.6t_c$$

and $$t_L = \frac{L^{0.8}(S+1)^{0.7}}{1900Y^{0.5}}$$

where L = Length to the watershed divide in feet

S = Potential maximum retention (in.) given by $S = \dfrac{1000}{CN} - 10$

CN = Curve number for the watershed

Y = Watershed slope in %.

The NRCS dimensionless unit hydrograph is given in tabular form in Table 4.9.1

In Table 4.9.1, t_p is the time to peak and q_p is the peak runoff rate. From these ratios, the unit hydrograph can be constructed.

The underlying theory of the NRCS method is developed below.

The NRCS method separates rainfall into direct runoff Q, initial abstraction Ia and actual retention F as shown in Figure 4.9.1.

Conceptually the following relationship is assumed:

$$\frac{F}{S} = \frac{Q}{P - Ia}$$

where F = Actual retention

S = Potential retention

Ia = Initial abstraction

Q = Runoff

$F = (P - Ia) - Q$.

Therefore, $$Q = \frac{(P - Ia)^2}{(P - Ia) + S}$$

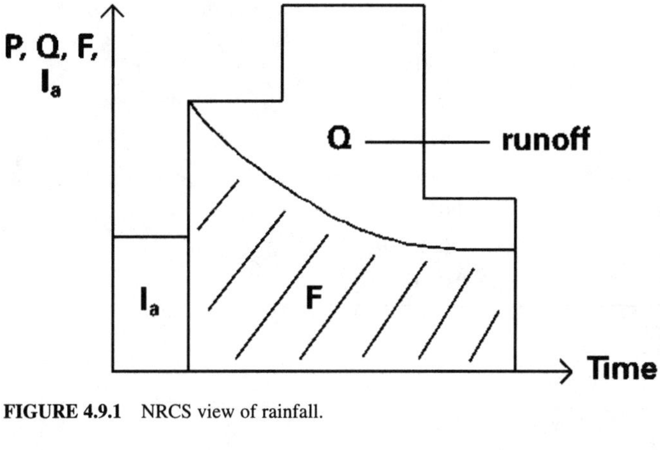

FIGURE 4.9.1 NRCS view of rainfall.

TABLE 4.9.1 NRCS Dimensionless Unit Hydrograph Ratios

Time ratio $= \dfrac{t}{t_p}$	Flowrate ratio $\dfrac{9}{q_p}$
0	0.000
0.1	0.030
0.2	0.100
0.3	0.190
0.4	0.310
0.5	0.470
0.6	0.660
0.7	0.820
0.8	0.930
0.9	0.990
1.0	1.000
1.1	0.990
1.2	0.930
1.3	0.860
1.4	0.780
1.5	0.680
1.6	0.560
1.7	0.460
1.8	0.390
1.9	0.330
2.0	0.280
2.2	0.207
2.4	0.147
2.6	0.107
2.8	0.077
3.0	0.055
3.2	0.040
3.4	0.029
3.6	0.021
3.8	0.015
4.0	0.011
4.5	0.005
5.0	0.000

Using the recommended NRCS relationship:

$$Ia = 0.25$$

Therefore,
$$Q = \frac{(P - 0.2S)^2}{P + 0.8S}$$

But,
$$S = \frac{1000}{CN - 10}$$

and it is assumed that $P \geq 0.2S$.

CHAPTER 5
GROUNDWATER HYDRAULICS

In this chapter, a survey of the major concepts in groundwater hydrology is presented. This is achieved by presenting basic groundwater hydrology principles and by providing application problems. The topics covered include: classification of groundwater, types of aquifers or groundwater formations, basic parameters in groundwater storage and movement, Darcy's Law, the Laplace equation for groundwater flow, flow nets and the application of flow nets, and groundwater well hydraulics including well field design.

5.0 SCOPE AND DEFINITIONS

The following sections present basic groundwater hydrology and hydraulics concepts, definitions and theory. In general, the material presented deals with groundwater storage and recharge groundwater flow, governing equations, basic design issues, flow nets and introductory well design guidelines. The guiding principle is to survey the subject matter to provide the reader with fundamental material to solve basic problems.

Groundwater or subsurface water studies focus on the waters below the surface of the earth. The waters move in the unsaturated (vadoze zone or zone of aeration) and saturated zones and move as deep percolation. The storage, movement, and distribution of subsurface waters and the engineering principles necessary to manage, extract, and protect groundwater supplies are of primary interest to the engineer and thus the focus of this chapter.

An aquifer is a geologic formation that is saturated with water and can transmit significant amounts of water. A significant portion of daily water use in the United States of America is drawn from groundwater sources. For example, in 1985, more than 14 states in obtained more than 60 percent of their water supplies from groundwater sources. Concerns about groundwater contamination have come into sharp focus recently, making the proper understanding of groundwater hydrology and hydraulics of great importance in most civil engineering programs and a vital subject area in water resources engineering. The engineering hydrologist must address both groundwater flow and contaminant fate and transport issues.

The following topics are the leading areas of concern:

1. Flow and contaminant transport processes
2. Field investigation methods
3. Flow and transport modeling
4. Remediation of contaminated groundwater sources.

5.1 GROUNDWATER SOURCES AND TYPES

This section provides basic and foundational information about groundwater sources and the subsurface environment. The types of aquifers and subsurface media are discussed to provide the context in which groundwater flow, storage, and contaminant fate and transport are examined.

Zones of Subsurface Water

The water table divides the *saturated zone* (pore spaces filled with water) from the *unsaturated zone* (zone of aeration). (Figure 5.1.2 delineates the zones that are encountered in the study of groundwater.)

Important terms are described below.

Gravitational water drains through the (unsaturated) subsurface under the influence of gravitational forces.

Capillary water rises above the water table due to surface tension forces. The area where capillary water is found is called the *capillary fringe*.

Hygroscopic water (in the form of a film) is held by adhesive forces onto the surface of soil grains. This water is bound somewhat to the grains and its availability to plants is limited by the suction pressure generated in the root system by plant species.

The unsaturated zone or *vadoze* is measured from the ground surface to the water table. In this zone, water is retained or adsorbed by adhesive (surface) forces and gravity pulls percolating water downward. The movement is retarded by both capillary (surface tension) and viscous forces.

The pore spaces in the saturated zone are filled with water. This zone is below the unsaturated zone (beneath the water table). Because groundwater yield depends on porosity, fine-grained media yield less water than coarse-grained media does. Thus, coarse-grain subsurface media serve as significant water-producing aquifers.

The *capillary fringe* extends from the water table to the limit of capillary rise, which depends on the size of the pore spaces. Capillary size in this zone can range from 250 mm for fine gravel to more than 2000 mm for silty subsurface media.

Aquifers are water-bearing formations that can yield significant quantities of water. The pore spaces in these formations are interconnected in such a way that water drains with relative ease. In view of this, alluvial deposits, which make up about 90 percent of developed aquifers, are the most productive types of aquifer media. Conversely, clays, though with high-

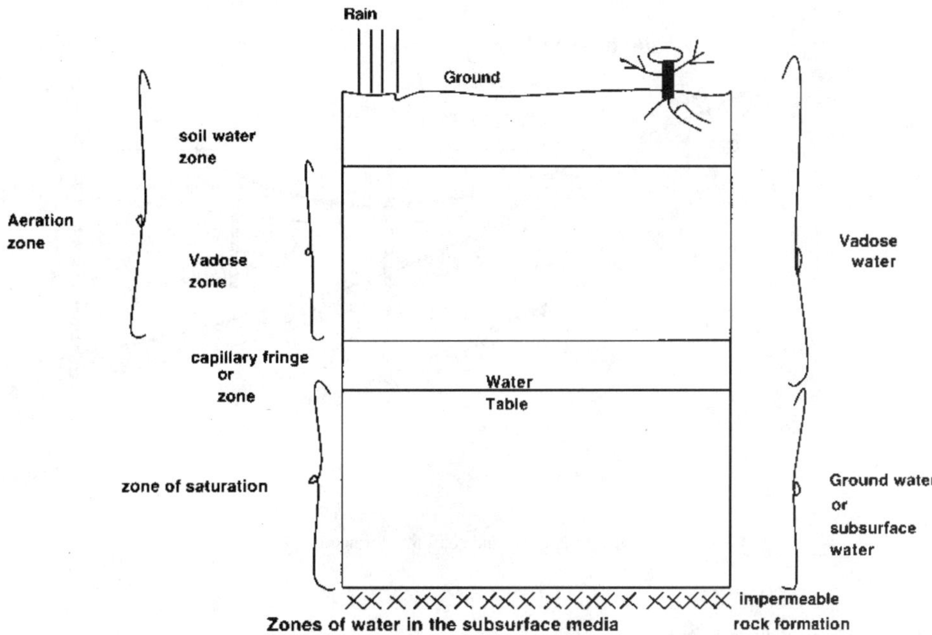

FIGURE 5.1.1 Zones of water in the subsurface environment.

level porosity, are poor yielding formations because of the relatively small pore spaces and their limited interconnectivity.

The two general classes of aquifers are *confined* and *unconfined* aquifers. Unconfined aquifers sometimes are called water table aquifers. In this case, the water table is virtually exposed to atmospheric pressure and thus, is shallow and may be used to extract water from it. Confined aquifers are overlain and underlain by confining layers. When the confining layer is semi-permeable, that is, a low permeability stratum, water can leak to the neighboring layers, and the confining unit is called an aquitard. A good example of an aquitard is a confining sandy clay unit. Impermeable confining units, such as clay or silt strata, are called *aquicludes*.

A *perched aquifer* is an unconfined aquifer in which the confining layer (below the aquifer) is a discontinuous and relatively impermeable layer that exists above a main body of subsurface water. In this case, the confining layer intercepts downward, moving water and forms a local body of groundwater. In Figure 5.1.2, the basic type of aquifers and confining layers are illustrated. When a well is developed to extract water from a groundwater source that is under pressure, an artesian or flowing well can develop because the pressure forces the water to rise above the level of the upper confining stratum.

5.2 CHARACTERIZATION OF SUBSURFACE MEDIA

This section provides descriptions and definitions of the subsurface media in which groundwater is stored or flows.

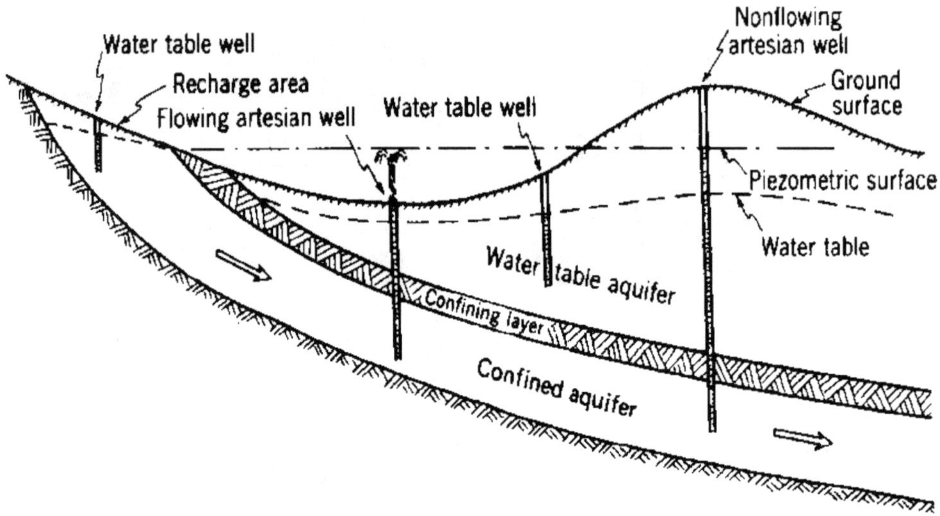

FIGURE 5.1.2 Types of aquifers and confining formations.

Porosity and Void Ratio

Porosity of a soil sample is the volumetric ratio of voids to the total sample volume. Figure 5.2.1 is a definitional figure for describing the soil sample

where V_v = Volume of voids (since pore spaces are filled with water or air/gas) and
$V_v = V_a + V_w$
V_w = Volume of water in the soil sample
V_a = Volume of the air/gas component
V_s = Volume of the solid component
V_t = Total volume of the sample such that
$V_t = V_v + V_s = V_a + V_w + V_s$
M_t = Total mass of the soil sample; so that
$M_t = M_s + M_w$ (Mass of air is negligible)
M_s = Mass of solid component of sample
M_w = Mass of water component of sample.

Therefore porosity is given by:

$$n = \frac{V_v}{V_t}$$

and the void ratio is given by

$$e = \frac{V_v}{V_s}$$

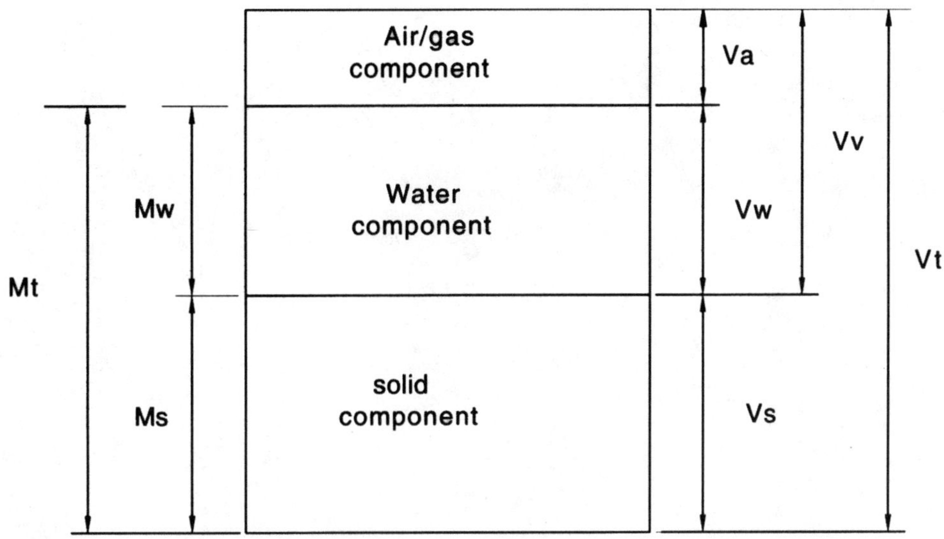

FIGURE 5.2.1 Components of a soil sample.

Thus $e \geq n$ and the relation between e and n can be found thus:

The dry or bulk density of a sample is given by

$$\rho_b = \frac{M_s}{V_t}$$

and so

$$V_t = \frac{M_s}{\rho_d}$$

The grain (solid) density of a sample is given by

$$\rho_s = \frac{M_s}{V_s}$$

and so

$$V_s = \frac{M_s}{\rho_s}$$

Since

$$n = \frac{V_v}{V_1}$$

and

$$V_v = V_t - V_s$$

therefore

$$n = \frac{V_t - V_s}{V_t} = 1 - \left(\frac{V_s}{V_t}\right)$$

$$n = 1 - \frac{\left(\frac{M_s}{\rho_s}\right)}{\left(\frac{M_s}{\rho_d}\right)}$$

thus

$$n = 1 - \left(\frac{\rho_d}{\rho_s}\right)$$

Also since

$$n = \frac{V_v}{V_t} \quad \text{and} \quad e = \frac{V_v}{V_s}$$

$$V_v = nV_t$$

$$V_s = V_t - V_v = V_t - nV_t = (1 - n)V_t$$

Thus

$$\begin{aligned} e &= \frac{V_v}{V_s} \\ &= \frac{V_v}{(1 - n)V_t} \\ &= \left(\frac{1}{1 - n}\right)(n) \\ &= \frac{n}{(1 - n)} \end{aligned}$$

and it also can be deduced that

$$n = \frac{e}{(1 + e)}.$$

The porosity of a water-bearing formation is an important indicator of its ability to store water but it is not the only parameter that indicates the yield from it. The combined indicators of the size of void spaces and the degree of interconnectivity between them determine the value of a water-bearing formation. Coarse sand is a good medium because void spaces are large and the degree of interconnection of the voids is high. On the other hand, clay is a poor aquifer medium because although the porosity is relatively high, the degree of interconnectivity of the void spaces is very low.

TABLE 5.2.1 Reference Porosity of Selected Materials

Material	Porosity (%)
Sand	
Coarse	39
Medium	39
Fine	43
Silt	46
Clay	42
Sandstone	
Medium	37
Fine	33
Claystone	43
Shale	6
Till	
Mostly sand	31
Mostly silt	34
Tuff	41
Weathered granite	45
Weathered gabbro	43
Basalt	17
Limestone	30
Dolomite	26
Dune sand	45
Loess	49
Schist	38
Siltstone	35
Peat	92
Pumice	85

Chin, page 463; Bedent (1999) page 20, T 2.1

Except for loess and pumice, the range of porosities for aquifer medium is between 6 percent and 46 percent. A representative value of 30 percent usually is assumed for aquifer media porosity unless fractures are present.

The Standard Soil Classification based on particle size is given in Table 5.2.2.

Specific Yield of an Unconfined Aquifer

The specific yield is the volume of water released from an unconfined or water table aquifer per unit horizontal surface area of the aquifer per unit head drop in the water table. Thus specific yield is given by

$$S_y = \left(\frac{1}{A}\right)\left(\frac{\Delta\forall}{\Delta h}\right)$$

TABLE 5.2.2 Soil Classification (Course—Fine)

Material	Grain size (mm)
Gravel	
Very coarse	32.0-64.0
Coarse	16.0-32.0
Medium	8.0-16.0
Fine	4.0-8.0
Very fine	2.0-4.0
Sand	
Very coarse	1.0-2.0
Coarse	0.5-1.0
Medium	0.25-0.5
Fine	0.125-0.25
Very fine	0.062-0.125
Silt	0.004-0.062
Clay	< 0.004

Bedent (1999) Page 21

where S_y = Specific yield of water table aquifer
A = Horizontal surface area of the soil formation
$\Delta\forall$ = Volume of water released
Δh = Change in water table elevation.

Since some moisture remains in the soil after the draining of water, $S_y \neq n$ (porosity).

The amount of water retained in the soil can be expressed by the specific retention S_r, which is defined as

$$S_r = n - S_y$$

where n = Porosity of the soil.

Reference for specific yield for some soils is given in Table 5.2.3.

Storage Coefficient of Storativity of a Confined Aquifer

The storage coefficient, or storativity (S), is defined (with the help of the specific storage, S_s) as:

$$S = S_s b$$

where b = Thickness of the confined aquifer.

The specific storage S_s is defined as the volume of water released (or recharged into) from the (elastic) storage of a confined aquifer, per unit change in pressure head of the aquifer. A confined aquifer is always saturated and the storage is elastic because decreasing the head

TABLE 5.2.3 Reference Values of Specific Yield

Material	Specific yield (%)
Gravel	
Fine	25
Medium	24
Coarse	23
Sand	
Fine	23
Medium	28
Coarse	27
Dune sand	38
Silt	8
Clay	3
Sandstone	
Fine grained	21
Medium grained	27
Limestone	14
Till	
Mostly silt	6
Mostly sand	16
Mostly gravel	16
Loess	18
Peat	44
Schist	26
Tuff	21
Siltstone	12

Todd (1980), page 130-Gupta

proportionally decreases the storage within the aquifer. The water in a confined aquifer is under pressure and the release of water from it causes pressure drop rather than (in contrast to unconfined aquifers) a drop in the volume (of the saturated medium).

For example, when water is pumped from a confined aquifer through a well, the pore pressure is reduced. This, in turn, leads to a slight compaction of the *frame* of the aquifer and an expansion (since water is elastic) of the water in pores. Hence, the total physical volume of the aquifer is unchanged, but water can be pumped out of it.

$$S_s = \frac{V_r}{V_t} \text{ for 1 unit drop in pressure head}$$

where V_r = Volume of water released from storage in a unit volume of the aquifer
V_t = Volume of the unit volume of the aquifer.

The analytical expression of S_s, which takes into account the elasticity of water in the pores and the elasticity of the frame of the aquifer or soil skeleton, is given by

$$S_s = \rho_w g(\alpha + n\beta)$$

where ρ_w = Density of water (in the pores)
β = Compressibility of water
n = Porosity of aquifer medium
α = Compressibility of the soil skeleton
g = Gravitational acceleration.

Another way to express this is to use the bulk modulus of water (E_w) and soil skeleton (E_s).

$$S_s = n\rho_w g\left[\left(\frac{1}{E_w}\right) + \left(\frac{1}{nE_s}\right)\right]$$

EXAMPLE 5.2.1

An unconfined aquifer has an average surface area of 10 hectares and the water table drops 2 m in 15 days. The aquifer soil is medium sand (with a specific yield of 22%). How much water has drained from the aquifer in million of liters, and what is the pumping rate in m^3/s*?*

Solution

$$1 \text{ hectare} = 10{,}000 \text{ m}^2$$

Therefore

$$\text{the average surface area} = 10 \times 10{,}000 \text{ m}^2 = 1 \times 10^5 \text{ m}^2$$

The volume of the aquifer from which the water was drained as the water table dropped = (Area)(drop in water table)

$$\text{Volume of aquifer} = 1 \times 10^5 \text{ m}^2(2\text{m}) = 2 \times 10^5 \text{ m}^3$$

The specifi yield, $S_y = 22$

Therefore

amount of water pumped (assuming a pumping well was installed)
= (S_y)(Volume of aquifer)

$$\text{Water pumped} = 0.22 \times 2 \times 10^5 = 0.44 \times 10^5 \text{ m}^3$$

Since 1 m^3 = 1000 liters

$$\text{Water pumped} = \frac{(1000 \times 0.44 \times 10^5 \text{ m}^3)}{1{,}000{,}000} = 44 \text{ million liters}$$

$$\text{The pumping rate} = \frac{\text{Volume pumped (m}^3)}{\text{time (s)}} = \frac{(0.44 \times 10^5 \text{ m}^3)}{(15)(24)(60)(60)} = 0.034 \text{ m}^3\text{/s or (34 l/s)}$$

Note: To help put this in perspective for those who still think in English units, this is about 539 gallons/min.

EXAMPLE 5.2.2

A 2-liter sample of soil from a confined aquifer was dried and found to have a mass of 3.36 kg.

In previous analysis, it was determined that the specific gravity of the aquifer grains (solids) was 2.4. Water in the aquifer is at 10°C (bulk modulus of 2.10×10^9 N/m^2) and density = 999.7 kg/m^3) and the compressibility (α) of the aquifer skeleton is 2×10^{-8} m^2/N). The average thickness of the aquifer is 50 m.

Determine:

1. *Porosity of the aquifer medium*
2. *Void ratio*
3. *Specific storage (S_s) of the aquifer*
4. *Storage coefficient (S) of the aquifer*

Solution

1. To determine the porosity (n):

$$n = 1 - \left(\frac{\rho_d}{\rho_s}\right)$$

where ρ_d = Bulk (dry) density of the soil
ρ_s = Grain density.

$$\rho_d = \frac{\text{Mass of soil}}{\text{Volume}}$$

$$= \frac{3.36 \text{ kg}}{\left(2 \text{ liters} \times \dfrac{1 \text{ m}^3}{1000 \text{ liters}}\right)} = 1680 \text{ kg/m}^3$$

$$\rho_s = \text{(Specific gravity)(density of water at 4°C)}$$
$$= (2.4 \times 1000 \text{ kg/m}^3) = 2400 \text{ kg/m}^3$$

therefore

$$n = 1 - \left(\frac{1680}{2400}\right) = 1 - 0.70 = 0.30$$

Therefore,

$$n = 30\%$$

2. Void ratio (e)

$$e = \frac{n}{1-n} = \frac{0.3}{(1-0.3)} = \left(\frac{0.3}{0.7}\right) = 0.429$$

Therefore,

$$e = 42.9\%$$

3. Specific storage (S_s)

$$S_s = \rho_w g(\alpha + n\beta)$$

All parameters are as defined previously in this section

$$n = 0.30 \quad \text{(found in part 1)}$$
$$\rho_w = 999.7 \text{ kg/m}^3 \quad \text{(density)}$$
$$\alpha = 2 \times 10^{-8} \text{ m}^2\text{/N} \quad \text{(compressibility)}$$
$$g = 9.81 \text{ m}^2\text{/s} \quad \text{(gravitational acceleration)}$$

$$\beta = \frac{1}{\text{Bulk modulus of water}}$$
$$= \frac{1}{2.10 \times 10^{-9} \dfrac{\text{N}}{\text{m}^2}} = 4.762 \times 10^{-10} \frac{\text{N}}{\text{m}^2}$$

Therefore

$$S_s = (999.7)(9.81)[(2 \times 10^{-8}) + (0.30)(4.762 \times 10^{-10})]$$
$$= 1.975 \times 10^{-4}) \text{ per m}$$

Approximately 2×10^{-4} m^{-1}

4. Storage coefficient (S)

$$S = S_s(b)$$

where b = Average aquifer thickness.

$$S = (1.975 \times 10^{-4} \text{ m}^{-1})(50 \text{ m}) = 9.9 \times 10^{-3}$$

Approximately 0.01.

Note: If a pressure head drop of 60 m occurs in this aquifer, the amount of water released (per unit area, that is, m^2) of the aquifer = $(S)(\Delta h) = 9.9 \times 10^{-3} \times 60$ m = 0.593 m^3 per m^2.

5.3 GROUNDWATER RECHARGE

Precipitation replenishes groundwater in an aquifer. Water can be withdrawn continuously from an aquifer only if it is recharged. If the aquifer is not replenished, it will eventually dry up.

The term *perennial yield* is used to describe the amount of water that can be extracted from an aquifer annually. This amount is limited by the level of recharge that can occur over the same period. Thus, the amount of recharge that the aquifer can obtain is the limit to which it can supply. This is limited physically by the aquifer's storage and transmission capacities.

To augment groundwater supplies, it is often necessary to induce recharge artificially. Artificial recharge methods include:

1. Recharge wells. Water is directed to a well so that recharge from the well enters the aquifer. Essentially, this is the reverse of pumping from a well.
2. Inducing flow from a stream. Induced infiltration can be designed to occur in a system where an aquifer is hydraulically interconnected with a stream. The major design feature is to create a hydraulic gradient that favors flow from the stream to the aquifer. This is sometimes achieved by rapidly pumping a well close to the connected stream, creating a favorable hydraulic gradient that causes the aquifer to be replenished.
3. Water spreading methods
   a. Basin method. Water is diverted to an artificial basin composed of a pervious material so that it seeps into the groundwater. A variant of this method is a flooding technique, in which water is flooded over a relatively flat terrain to increase infiltration.
   b. Ditch method. In this variation of the basin method, flat-bottomed shallow ditches are constructed so that the spacing between ditches is small.
   c. River-bed method. This method often is used in alluvial streams where water readily infiltrates into the underlying river bed stratum. Measures, however, are taken to enhance the infiltration by widening the river bed and by building check dams and dikes.

5.4 HYDRAULICS OF GROUNDWATER

In this section, the parameters that are important in describing groundwater movement are discussed. This section provides the background information essential for understanding most of the other sections that follow.

Coefficient of Permeability or Hydraulic Conductivity

In the equation of flow through a porous medium, the flow rate (Q) is directly proportional to the energy (head) gradient ($\Delta h \backslash L$) and the area (A) through which the flow is occurring. Therefore, the following relationship exists (Δh is the drop in head along the path of length, L):

$$Q \propto A \frac{\Delta h}{L}$$

The constant of proportionality is called the coefficient of permeability or the hydraulic conductivity (K).

Thus

$$Q = KA\frac{\Delta h}{L} \qquad i = \frac{\Delta h}{L}$$

The term K is a measure of both fluid (flowing) properties and the properties of the porous medium with respect to the transmission of water through the system. The following relationship, however, provides a means for accounting for the ability of the medium only to transmit flow:

$$K = k\frac{\gamma}{\mu}$$

where K = Hydraulic conductivity (ft/day, m/day)
k = Intrinsic permeability (m^2, ft^2)
γ = Specific weight of the liquid (N/m^3, $1b/ft^3$)
μ = Coefficient of viscosity (dynamic viscosity) of the liquid $\left(\frac{N \cdot s}{m^2}, \frac{1b \cdot s}{ft^2}\right)$

Therefore,

$$K = k\frac{\gamma}{\mu} = \left(\frac{K}{g}\right)\left(\frac{\mu}{\rho}\right) = \frac{K\nu}{g}$$

where ν is the kinematic viscosity of the liquid.

Let q be the discharge per unit area, then

$$q = K\frac{\Delta h}{L} \text{ where } \frac{\Delta h}{L} = i$$

Therefore

$$q = Ki$$

and

$$K = \frac{q}{i}$$

and the intrinsic permeability can be defined as

$$K = \frac{K\nu}{g} = \frac{q\nu}{ig}$$

Representative values of K are provided in Table 5.4.1.

It should be noted that K shows a wide range of values (from 0.0002 to 450). In fact, K can vary by more than 12 orders of magnitude. Variations of three orders of magnitude can be encountered at a site. When modeling regional systems, variations of six orders of magnitude or more can be expected.

It can be shown by dimensional analysis (using the Buckingham pi theorem) that hydraulic conductivity can be separated into two parts—one accounting for the geometry of the solid

TABLE 5.4.1 Reference Values of *K* [m/day]

Material	Hydraulic Conductivity [m/day]
Gravel	
Fine	450
Medium	270
Coarse	150
Sand	
Fine	2.5
Medium	12
Coarse	45
Dune Sand	20
Silt	0.08
Clay	0.0002
Basalt	0.01
Sandstone	
Fine grained	0.2
Medium grained	3.1
Limestone	0.94
Slate	0.00008
Till	
Mostly sand	0.49
Mostly gravel	30
Loess	0.08
Peat	5.7
Schist	0.2
Tuff	0.2
Weathered gabbro	0.2
Weathered granite	1.4

Todd, 1980 Gupta 135

matrix and the other term (part) related to fluid properties. The term that is related to the geometry of the solid matrix is equal to the intrinsic permeability.

Thus, intrinsic permeability accounts solely for the porous medium through which the liquid flows. A number of equations have been proposed for estimating intrinsic permeability based on physical properties (usually pore size, mean grain diameter, shape factors, etc.) of an aquifer matrix. Intrinsic permeability is often measured in darcys.

$$1 \text{ darcy} = 0.987 \times 10^{-6} \text{ m}^2$$

Because hydraulic conductivity K depends on the specific weight γ and the dynamic viscosity (μ) of the fluid, and γ and μ are influenced by the temperature of the fluid, it is important to exercise care in estimating K.

TABLE 5.4.2 Conversion Factors for Hydraulic Conductivity, K

	Equivalent Values		
$1\ \frac{\text{gal}}{\text{day-ft}^2} =$	$0.0408\ \frac{\text{m}}{\text{day}}$	$0.134\ \frac{\text{ft}}{\text{day}}$	$4.72 \times 10^{-7}\ \frac{\text{m}}{\text{s}}$
$1\ \frac{\text{ft}}{\text{day}} =$	$0.305\ \frac{\text{m}}{\text{day}}$	$7.481\ \frac{\text{gal}}{\text{day-ft}^2}$	$3.53 \times 10^{-6}\ \frac{\text{m}}{\text{s}}$
$1\ \frac{\text{cm}}{\text{s}} =$	$864\ \frac{\text{m}}{\text{day}}$	$2835\ \frac{\text{ft}}{\text{day}}$	$21{,}200\ \frac{\text{gal}}{\text{day-ft}^2}$
$1\ \frac{\text{m}}{\text{day}} =$	$1.11 \times 10^{-5}\ \frac{\text{m}}{\text{s}}$	$3.281\ \frac{\text{ft}}{\text{day}}$	$24.5\ \frac{\text{gal}}{\text{day-ft}^2}$

Table 5.4.3 presents a classification of hydraulic conductivity values.

Transmissivity T. Transmissivity is the measure of an aquifer's capacity to transmit water through its entire depth (thickness) through which water flows. It also is defined as the product of K and the saturated thickness b of an aquifer.

$$T = Kb \quad \text{(for uniformly thick aquifer)}$$

where K = Average hydraulic conductivity
b = Thickness of the aquifer

It is evident from the foregoing that it is necessary to calculate average K in order to find the transmissivity of an aquifer.

Calculation of Average K. If the hydraulic conductivity varies smoothly with depth,

$$K(z) = f(z)\text{(a function of depth, } z\text{)}$$

TABLE 5.4.3 Ranges of Hydraulic Conductivity Values

Class of K	Range of K values [m/day]	Unconsolidated soil	Consolidated material
Very High	Greater than 1000	Clean gravel	Basalt cavernous limestone and dolomite
High	10-1000	Clean sand, sand, gravel	Clean sandstone, fracture igneous, metamorphic rocks
Moderate	0.01-10	Fine sand	Laminated shale, sandstone, mudstone
Low	0.0001-0.01	Silt, clay, mixtures of sand, silt, and clay	Massive igneous and metamorphic rocks
Very Low	Less than 0.0001	Massive clay	—

Then average K is given by

$$\bar{K} = \frac{1}{b}\int_{n}^{b} f(z)\,dz$$

However, if the medium is stratified, the average of K depends on the direction of flow that is to be considered.

Case 1—Flow parallel to stratification

$$\bar{K}_{\text{para}} = \frac{1}{\sum_{i=1}^{n} b_i} \sum_{i=1}^{n} K_i b_i$$

where $\bar{K}_{\text{para}}$ is the average K parallel to the stratification.

Case 2—Flow perpendicular to stratification

$$\bar{K}_{\text{norm}} = \frac{\sum_{i=1}^{n} w_i}{\sum_{i=1}^{n} \frac{w_i}{K_i}}$$

where $\bar{K}_{\text{norm}}$ is the average K normal to the direction of stratification.

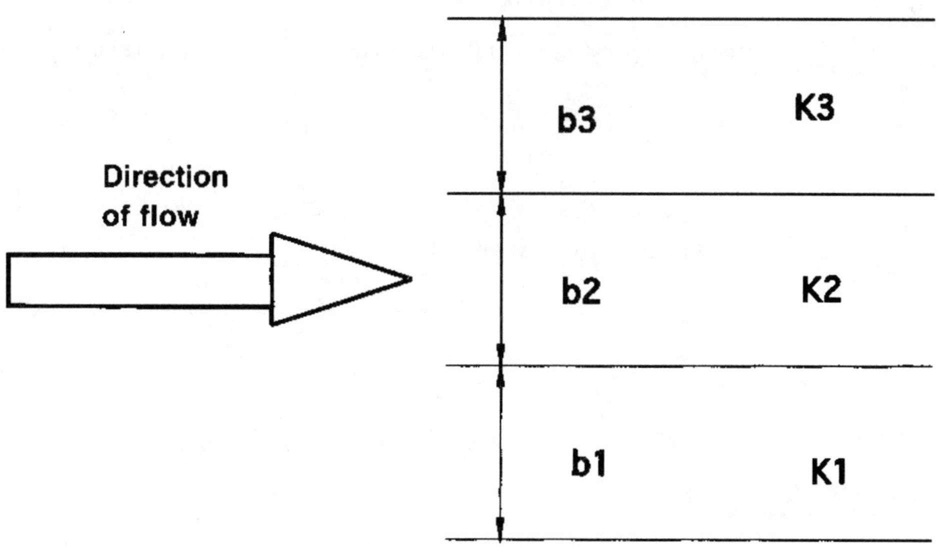

FIGURE 5.4.1 Flow and orientation of stratification (parallel case).

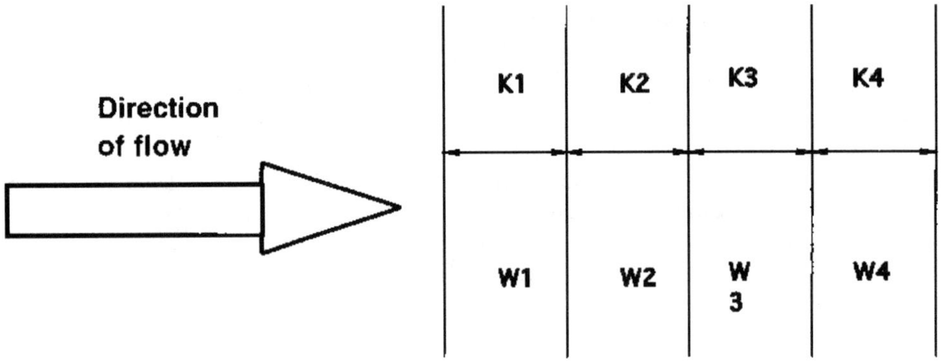

FIGURE 5.4.2 Flow and orientation of stratification (perpendicular case).

Leaky Aquifer Characteristics

1. Leakance or coefficient of leakage

   Leaky aquifers are aquifers in which the confining layer can transmit water. When an aquifer is confined by an impermeable layer, it is said to be confined by an *aquifuge*. An *aquitard* is a low permeability confining layer that can store groundwater and transmit it slowly from one aquifer to another. This is a leaky confining layer. The ability of a confining layer to transmit flow vertically is the leakance *Le*.

   Thus

$$Le = \frac{K_L}{b_L}$$

where K_L = The K for the leaky layer
b_L = Thickness

The retardance coefficient *R*, which indicates the areal distribution of leakage, is given by:

$$R = \frac{K}{\left(\frac{K_L}{b_L}\right)}$$

The leakage factor *B* is given by

$$B = \sqrt{\frac{K}{\left(\frac{K_L}{b_L}\right)}} = \sqrt{\frac{Kb}{Le}}$$

5.5 DARCY'S LAW

The fundamental equation of groundwater flow (hydraulics) is expressed as an equation called Darcy's Law. This law is attributed to Henri Philbert Gaspard Darcy (1803-1858),

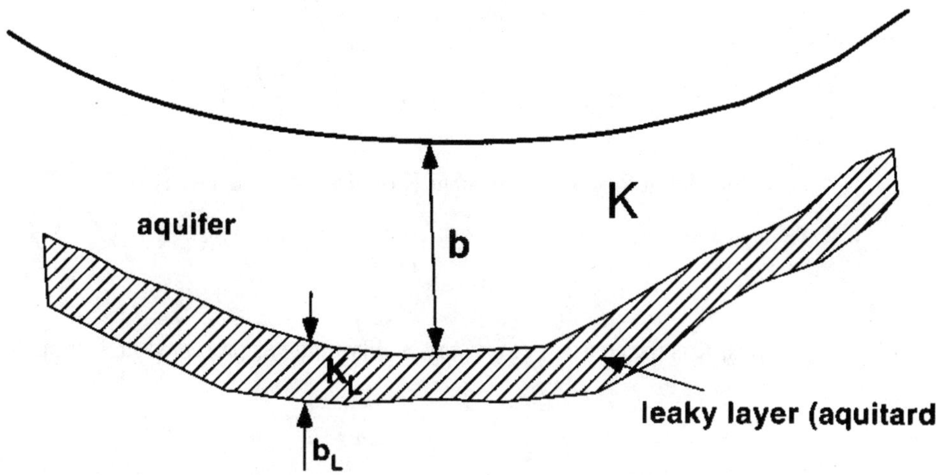

FIGURE 5.4.3 Leaky aquifer configuration.

a native of Dijon, France, who contained in an 1856 treatise his findings from experiments conducted in the design of filters for the public waterworks of Dijon.

Darcy found the following relationships with respect to Figure 5.5.1

$Q \, \alpha \, A$

$Q \, \alpha \, h_1 - h_2$ (drop in head)

$Q \, \frac{1}{\alpha} \, L$ (length of flow path)

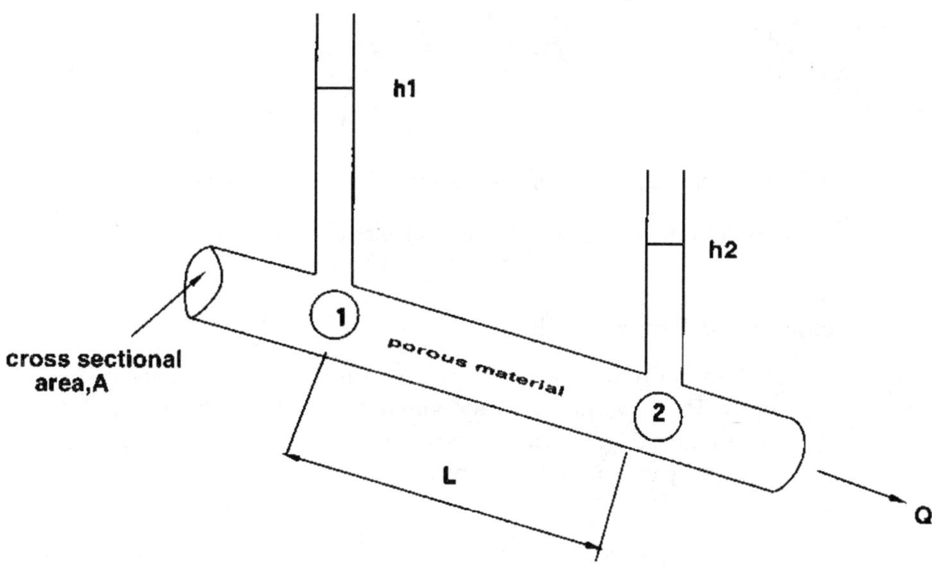

FIGURE 5.5.1 Basic setup for Darcy's experiment.

Combining the above relationships:

$$Q \, \alpha \, A\left(\frac{h_1 - h_2}{L}\right)$$

he found that for a given constant K (hydraulic conductivity),

$$Q = \mathrm{K}A\left(\frac{h_1 - h_2}{L}\right)$$

Thus, K is the constant of proportionality that was presented in Section 5.4.

Let

$$\Delta h = h_1 - h_2$$

and

$$\frac{\Delta h}{L} = i$$

Then

$$Q = \mathrm{K}Ai$$

Let

$$\frac{Q}{A} = q \quad \text{(specific discharge)}$$

Then

$$q = \frac{Q}{A} = -\mathrm{K}\frac{\Delta h}{L}$$

where the negative sign indicates that flow is in the direction of the drop in head.

The specific discharge q is the discharge per unit cross-sectional area.

Darcy Velocity and Seepage Velocity

The area of flow through a porous medium is not the area given by A as described previously. The flow path takes a tortuous route, traversing the pore spaces in the medium. Consider Figure 5.5.2 in which a cross section of the porous medium is depicted. Previously the specific discharge, q was described mathematically by

$$q = -\mathrm{K}\frac{\Delta h}{L}$$

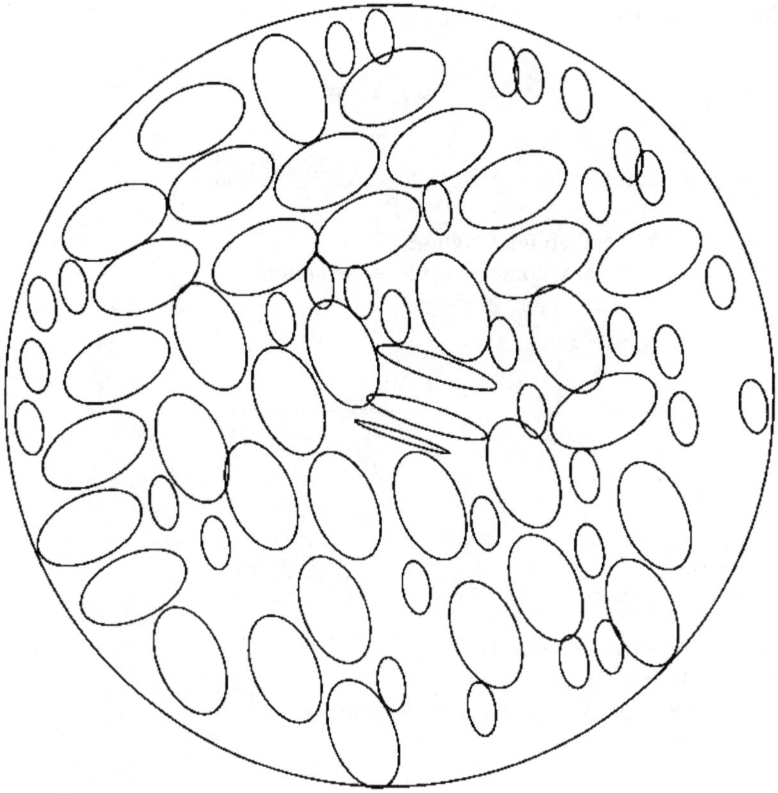

FIGURE 5.5.2 Cross-section of porous medium.

We can define a discharge velocity V such that $q = V$ (regular definition of velocity) since

$$V = \frac{Q}{A}$$

Thus

$$AV = qA$$

Let V_v be the seepage velocity through the medium and A_v be the actual area of voids over a cross section as shown in Figure 5.5.2, then

$$Q = AV = Aq = A_v V_v$$

Therefore

$$V_v = \frac{VA}{A_v}$$

Multiplying both sides by the length of the medium,

$$V_v L = \frac{AL}{A_v}$$

$$V_v = V\frac{AL}{A_v L} = V\frac{\forall}{\forall_v}$$

where $\forall$ = Total column volume
$\forall_v$ = Total volume of voids in the column

But

$$\frac{\forall_v}{\forall} = n - \text{(porosity)}$$

Therefore

$$V_v = \frac{V}{n}$$

where V = Darcy velocity
V_v = Seepage velocity
n = Porosity of the aquifer medium

Note that

$$V_v > V \quad \text{since } 0 < n < 1.0$$

Validity of Darcy's Law

Darcy's equation is valid only when the discharge velocity is proportional to the hydraulic gradient $\Delta h/L$. When flow is laminar (as defined by the Reynolds number), velocity has a linear relationship with $\frac{\Delta h}{L}$.

For a porous medium, Reynolds number is definedν as:

$$Re = \frac{Vd_{10}}{\nu}$$

where V = Darcy velocity
ν = Kinematic viscosity of the liquid (water)
d_{10} = Effective grain size (10% of materials are smaller than d_{10}).

When the Reynolds number is less than 1, as in this situation, laminar flow occurs. However, the limit $Re \leq 10$ is used.

In addition, Darcy's Law applies only when the porous medium is fully saturated.

5.6 GENERAL GROUNDWATER FLOW EQUATIONS

In this section, a generalized version of Darcy's (empirical) one-dimensional equation is presented in a three-dimensional generalized form. Secondly, the general equations for groundwater flow, including the special case of the Laplace equation for groundwater movement and storage, are presented as well.

General Form of Darcy's Law

The total head, h, at a point in a flow is given by

$$h = \frac{V^2}{2g} + \frac{P}{8} + z$$

where $\frac{V^2}{2g}$ = velocity head, but V is interpreted as the specific discharge, q

g = gravitational acceleration

z = elevation or position with respect to a given datum

P = pressure

γ = specific weight of the liquid (water)

The velocity head $\frac{V^2}{2g}$ is ignored because of the relatively low magnitude of the specific discharge. Thus,

$$h = z + \frac{P}{\gamma}$$

But

$$V = q = -\mathrm{K}i = -\mathrm{K}\frac{\partial h}{\partial \rho}$$

Thus in 3-dimensions:

$$V_x = -\mathrm{K}_x \times \frac{\partial h}{\partial x}$$

$$V_y = -\mathrm{K}_y \times \frac{\partial h}{\partial y}$$

$$V_z = -\mathrm{K}_z \times \frac{\partial h}{\partial z}$$

In vector notation

$$\mathbf{V} = V_x\mathbf{i} + V_y\mathbf{j} + V_z\mathbf{K}$$

Thus

$$\mathbf{V} = -\left[-\mathrm{K}_x\frac{\partial h}{\partial x}\mathbf{i} + \mathrm{K}_y\frac{\partial h}{\partial y}\mathbf{j} + \mathrm{K}_z\frac{\partial h}{\partial z}\mathbf{K}\right]$$

If the medium is isotropic then K values are the same in all directions

$$\mathrm{K} = \mathrm{K}_x = \mathrm{K}_y = \mathrm{K}_z$$

$$\mathbf{V} = -\mathrm{K}\left[\frac{\partial h}{\partial x}\mathbf{i} + \frac{\partial h}{\partial y}\mathbf{j} + \frac{\partial h}{\partial z}\mathbf{K}\right]$$

or

$$\mathbf{V} = -\mathrm{K}\nabla h$$

where ∇h is the gradient vector of head (h).

Groundwater Flow Equations for Confined and Unconfined Aquifers

For a general flow, consider a unit volume of porous media. This is referred to as an elemental control volume as shown in Figure 5.6.1.

Applying the law of conservation of mass for the control volume (CV):

(Mass flux into CV − Mass flux out of CV) = change in mass per unit time.

This is the continuity condition.

This above is written as:

$$\left[\frac{\partial}{\partial x}(\rho V_x) + \frac{\partial}{\partial y}(\rho V_y) + \frac{\partial}{\partial z}(PV_z)\right]\Delta x\Delta y\Delta z = -\frac{\partial M}{\partial t}$$

The negative sign on the right hand side is to account for the assumed convention that mass (M) depletion results in net positive flux.

Darcy's Law provides definition of V:

$$V_x = -\mathrm{K}_x\frac{\partial h}{\partial x},\quad V_y = -\mathrm{K}_x\frac{\partial h}{\partial y},\quad V_z = -\mathrm{K}_x\frac{\partial h}{\partial z}$$

Substituting into the continuity equation, canceling the negative sign on both sides and noting that

$$\frac{\partial \rho Vx}{\partial x} = \rho\frac{\partial Vx}{\partial x} + Vx\frac{\partial \rho}{\partial x} \quad \text{(also for } y \text{ and } z\text{)}$$

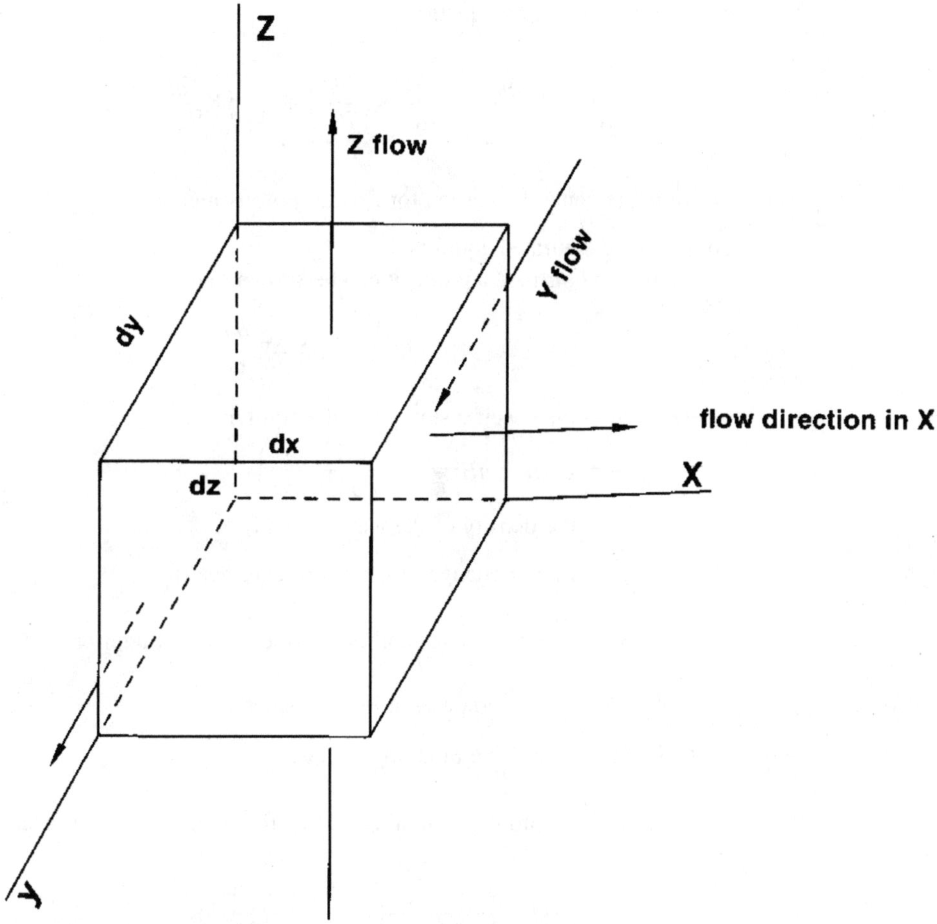

FIGURE 5.6.1 Elemental Control Volume

but

$$\frac{\partial \rho}{\partial x} \quad \text{and} \quad \frac{\partial \rho}{\partial y} \quad \text{and} \quad \frac{\partial \rho}{\partial z}$$

are all negligible in non-stratified flow conditions when compared to the specific velocity (V) change in all directions, it simplifies to

$$\frac{\partial \rho Vx}{\partial x} \cong \rho \frac{\partial Vx}{\partial x} \quad \text{(same for } y \text{ and } z \text{ directions)}$$

Letting

$$\Delta \forall = \Delta x \Delta y \Delta z$$

and simplifying continuity equation:

$$\left[\frac{\partial}{\partial x}\left(\mathrm{K}_x\frac{\partial h}{\partial x}\right)+\frac{\partial}{\partial y}\left(\mathrm{K}_y\frac{\partial h}{\partial y}\right)+\frac{\partial}{\partial z}\left(\mathrm{K}_z\frac{\partial h}{\partial z}\right)\right]\rho\Delta\forall=\frac{\partial M}{\partial t}$$

The above is the general equation for flow in porous media.

a) For confined saturated aquifers
In a saturated confined aquifer, it can be shown that

$$\frac{\partial M}{\partial t}=\rho S_s\Delta\forall\frac{\partial h}{\partial t}$$

where S_s = the specific storage of the aquifer

$$=\rho g(\alpha+n\beta)=\rho g\left(\frac{1}{Ea}+\frac{n}{Ew}\right)$$

ρ = the density of the water (liquid)

$\frac{\partial h}{\partial t}$ = rate of change of head (pressure head)

$\alpha=\frac{1}{Ea}$ = compressibility of solid matrix of aquifer

$\beta=\frac{1}{Ew}$ = compressibility of water

E = respective modulus of elasticity

Substituting $\frac{\partial M}{\partial t}$ into the general flow equation and dividing both sides by $\rho\Delta\forall$

$$\frac{\partial}{\partial x}\left(\mathrm{K}_x\frac{\partial h}{\partial x}\right)+\frac{\partial}{\partial y}\left(\mathrm{K}_y\frac{\partial h}{\partial y}\right)+\frac{\partial}{\partial z}\left(\mathrm{K}_z\frac{\partial h}{\partial z}\right)=S_s\frac{\partial h}{\partial t}$$

This is the groundwater flow equation for a non-homogeneous, anisotropic confined, saturated aquifer. The right hand side represents the water produced by the aquifer. A number of special conditions can be identified providing simplified relationships:

i) Homogeneous anisotropic case:
Here K_x, K_y, K_z do not depend on position in the x, y, z directions. For example, K_x is the K anywhere in the x-direction. The flow equation becomes:

$$\mathrm{K}_x\frac{\partial^2 h}{\partial x^2}+\mathrm{K}_y\frac{\partial^2 h}{\partial y^2}+\mathrm{K}_z\frac{\partial^2 h}{\partial z^2}=S_s\frac{\partial h}{\partial t}$$

ii) For homogeneous, isotropic case:

$$\mathrm{K}_x=\mathrm{K}_y=\mathrm{K}_z=\mathrm{K}$$

$$\frac{\partial^2 h}{\partial x^2}+\frac{\partial^2 h}{\partial y^2}+\frac{\partial^2 h}{\partial z^2}=\frac{S_s}{\mathrm{K}}\frac{\partial h}{\partial t}$$

iii) For steady flow in homogeneous, isotropic aquifer:

$$\frac{\partial^2 h}{\partial x^2} + \frac{\partial^2 h}{\partial y^2} + \frac{\partial^2 h}{\partial z^2} = 0$$

This is Laplace's equation and can be rewritten as

$$\nabla^2 h = 0$$

iv) For a horizontal confined aquifer of thickness b:

The storativity (storage coefficient) $S = bS_s$

For case (ii) above:

$$\frac{\partial^2 h}{\partial x^2} + \frac{\partial^2 h}{\partial y^2} + \frac{\partial^2 h}{\partial z^2} = \frac{S_s b}{\mathrm{K}} \frac{\partial h}{\partial t}$$

But

$$S_s b = S \qquad \text{and} \qquad \mathrm{K}b = T \quad \text{(Transmissivity)}$$

Therefore

$$\frac{\partial^2 h}{\partial x^2} + \frac{\partial^2 h}{\partial y^2} + \frac{\partial^2 h}{\partial z^2} = \frac{S}{T} \frac{\partial h}{\partial t}$$

v) Incorporating source or sink term [$W(x, y, z, t)$]. The source/sink term allows for the consideration of recharge or pumping (discharge) from an aquifer and the general equation for a confined aquifer becomes:

$$\frac{\partial}{\partial x}\left(\mathrm{K}_x \frac{\partial h}{\partial x}\right) + \frac{\partial}{\partial y}\left(\mathrm{K}_y \frac{\partial h}{\partial y}\right) + \frac{\partial}{\partial z}\left(\mathrm{K}_z \frac{\partial h}{\partial z}\right) \times \pm W(x, y, z, t) = S_s \frac{\partial h}{\partial t}$$

b) For unconfined saturated aquifers
In an unconfined aquifer, horizontal flow is assumed so that there is no vertical flow. In such a case, the problem space shrinks to a z-dimensional problem. Hence only the x and y directions are considered:
Since the density ρ is constant:

$$\left[\frac{\partial}{\partial x}\left(\mathrm{K}_x \frac{\partial h}{\partial x}\right) + \frac{\partial}{\partial y}\left(\mathrm{K}_y \frac{\partial h}{\partial y}\right)\right] \Delta x \Delta y = \frac{1}{\rho} \frac{\partial M}{\partial t}$$

$$= \frac{\partial V_w}{\partial t}$$

For an unconfined aquifer

$$V_w = Sydh\Delta x \Delta y$$

where Sy = specific yield of the aquifer

and

$$\frac{\partial V_w}{\partial t} = Sy \frac{\partial h}{\partial t} \Delta x \Delta y$$

Finally:

$$\frac{\partial}{\partial x}\left(K_x \frac{\partial h}{\partial x}\right) + \frac{\partial}{\partial y}\left(K_y \frac{\partial h}{\partial y}\right) = Sy \frac{\partial h}{\partial t}$$

i) For a case where the change in water table elevation is small compared to the unconfined aquifer thickness, *b*:

$$K_x \frac{\partial^2 h}{\partial x^2} + K_y \frac{\partial^2 h}{\partial y^2} = \frac{Sy}{b} \frac{\partial h}{\partial t}$$

ii) For a homogeneous, isotropic medium

$$\frac{\partial^2 h}{\partial x^2} + \frac{\partial^2 h}{\partial y^2} = \frac{Sy}{Kb} \frac{\partial h}{\partial t} = \frac{Sy}{T} \frac{\partial h}{\partial t}$$

5.7 FLOW NETS IN ISOTROPIC MEDIA

Flow nets is a graphical method for solving the groundwater flow equation discussed in the previous section. Flow nets are made up of streamlines (which are sometimes called flow lines) and equipotential lines. The streamlines and equipotential lines are constructed so that they cross at right angles. Flow lines are drawn to indicate the direction of flow and equipotential lines are drawn based on the observed level of water in wells (penetrating the aquifer) and any water table elevations observed or measured. Flow nets usually form approximately square grids.

Assumptions for Flow Nets in Isotropic Media

i) The aquifer is homogeneous (K is same in all directions)
ii) Flow is steady—no change in the heads with time
iii) The aquifer is fully saturated (assumption in Section 5.6)
iv) Incompressible flow for porous media matrix and the water in the pores
v) The flow is such that Darcy's Law is void ($Re < 10$) and thus laminar
vi) Necessary boundary conditions are known

Procedure for Developing Flow Nets

In summary, flow nets are made up of equipotential lines which are lines of equal hydraulic head and streamlines (flow lines) which are lines describing the flow path of water (liquid) particles. The equipotential lines and flow lines together represent the geometric solution of the two-dimensional equation of groundwater flow when the flow is assumed to be steady and homogeneous isotropic conditions are assumed.

Considering Figure 5.7.1, and the flow through one of the flow channels, the discharge per unit width of the aquifer (as shown)

$$\Delta q = \text{K(Area)(hydraulic gradient)}$$

$$= K(d_b \times i)\left(\frac{\Delta h_1}{dl}\right)$$

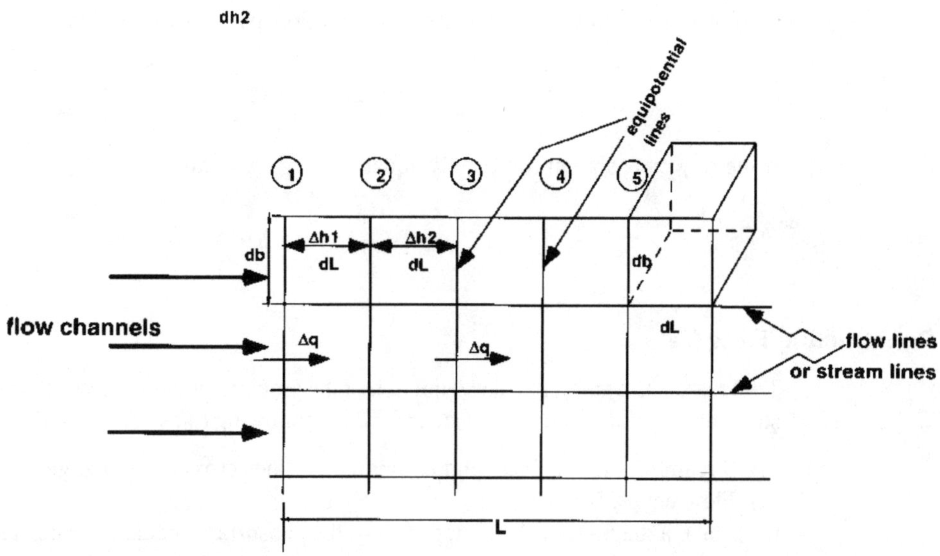

FIGURE 5.7.1 Definition sketch for flow nets.

where K = the hydraulic conductivity
d_b = the slice of aquifer thickness as shown in Figure 5.7.1
Δh_1 = the drop in head between equipotential (1) and (2)
dl = the distance between equipotential (1) and (2)

Flow per unit width is

$$\Delta q = \mathrm{K}(d_b \times 1)\frac{\Delta h_1}{dl} = \mathrm{K}(d_b \times 1)\frac{\Delta h_2}{dl}$$

Thus the flow nets will be constructed so that

$$\Delta h_1 = \Delta h_2 = \cdots = \Delta h_n$$

If h is the total head loss along the flow line (that is between the first and last equipotential lines)

Therefore
$$\Delta h = \frac{h}{n_e}$$

where n_e = number of equipotential drops

Therefore flow through one flow channel is given by

$$\Delta q = \mathrm{K}\frac{d_b}{dl}\frac{h}{n_e}$$

If there are n_f flow channels, then the total flow per unit width of the saturated aquifer is:

$$q = n_f \Delta q = \frac{n_f}{n_e} K \frac{d_b}{dl} h$$

If the flow nets are approximately square then $d_b = dl$ and

$$q = \frac{n_f}{n_e} K h$$

Constructing Flow Nets

To construct flow nets, a scaled depiction of the problem situation must be provided. The construction is a trial-and-error procedure and the following rules help to guide the construction:

i) Streamlines (flow lines) and equipotential lines cross at right angles
ii) Flow net grids must be approximately square
iii) Make a diagram of the boundaries of the subsurface media to scale. The x and y scales must be the same
iv) Identify all known equipotential and flow line stipulations
v) All impermeable boundaries must be shown as flow lines
vi) The soil/water interface at both upstream and downstream sections of a structure are equipotential lines

An example is shown in Figure 5.7.2.

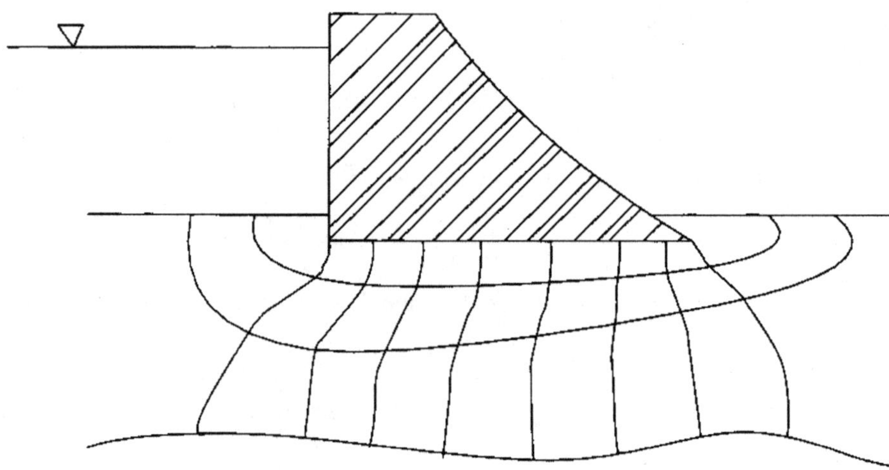

FIGURE 5.7.2 Example flow net construction

APPENDIX A
GROUND WATER

Source: U.S. Department of the Interior, Bureau of Reclamation

(1) One - 6mm (¼in) stainless steel stop cock.
(2) One - 1.2m (4ft) length of 16mm (⅝in) i.d. rubber hose.
(3) One - 50mm (2in) dia. ink bottle.
(4) One - 3 holed No. 8 rubber stop.
(5) One - 19mm (¾in) hose coupling.
(6) One - 1200mm (48in) length of 2mm i.d. glass tubing.
(7) One - 1125mm (45in) length of stainless steel strip with graduations which give readings in m (ft) of water.
(8) One - 100mm (4in) length of 6mm (¼in) o.d. stainless steel tubing with fittings.
(9) One - 8mm ($^{5}/_{16}$in) stainless steel stop cock.
(10) One - 100mm (4in) length of 8mm ($^{5}/_{16}$in) o.d. stainless steel or plastic tubing with fitting.

Assorted lumber (marine plywood)
Assorted 3mm (⅛in) bolts with nuts.

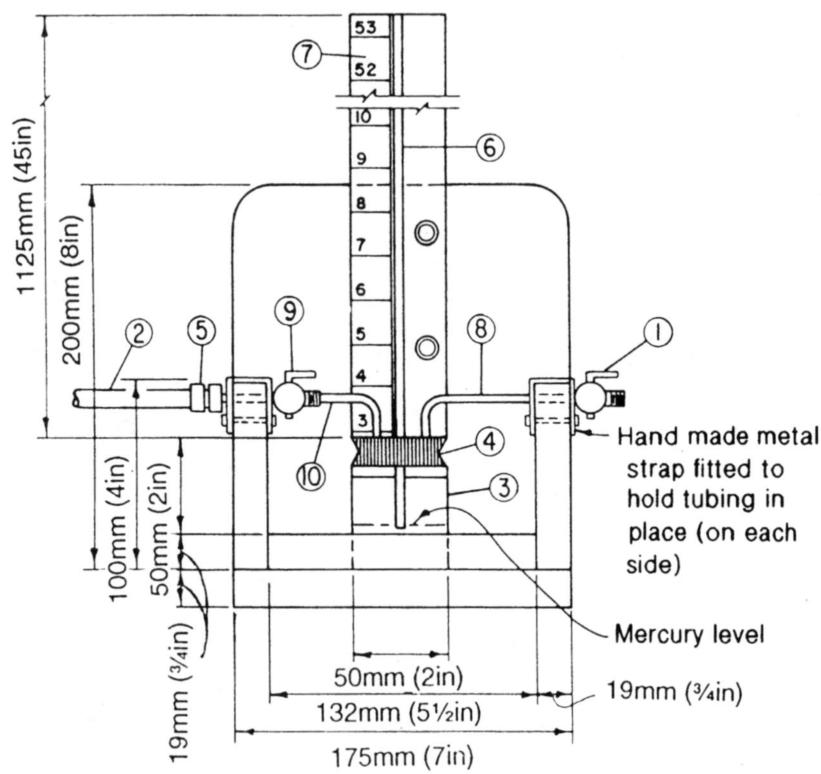

(After S. W. Lohman)

Mercury manometer for measuring artesian heads.

FIGURE A-1 Mercury manometer for measuring artesian heads.

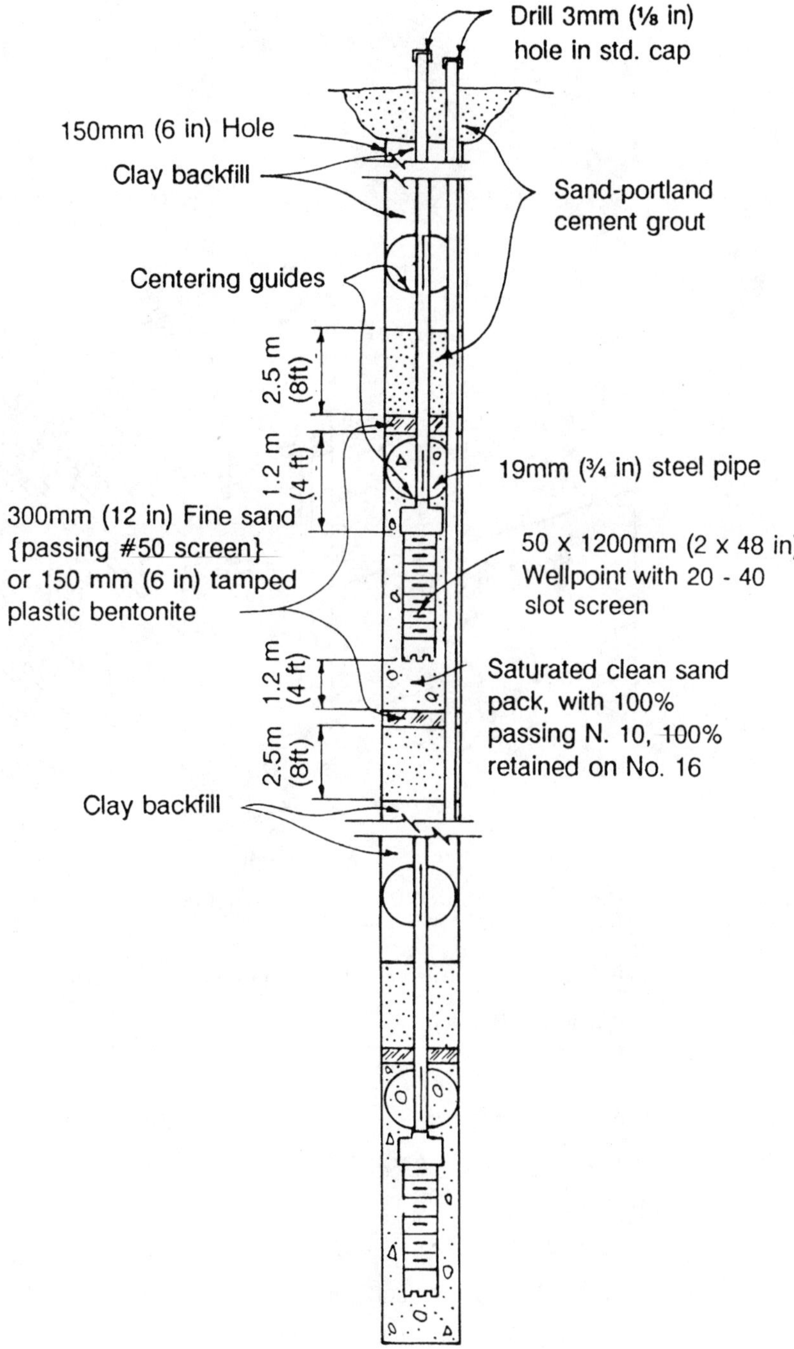

Typical dual piezometer installation.

FIGURE A-2 Typical dual piezometer installation.

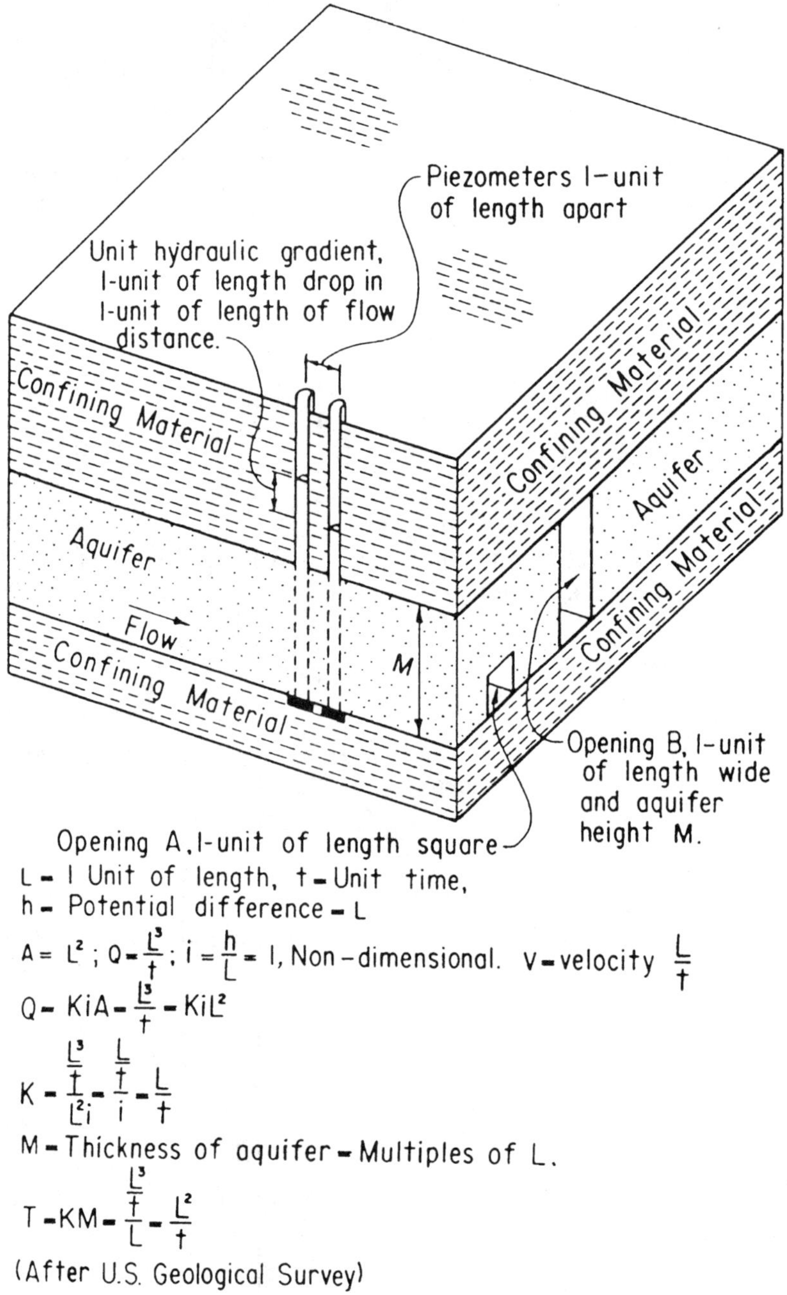

Illustration of hydraulic conductivity and transmissivity.

FIGURE A-3 Illustration of hydraulic conductivity and transmissivity.

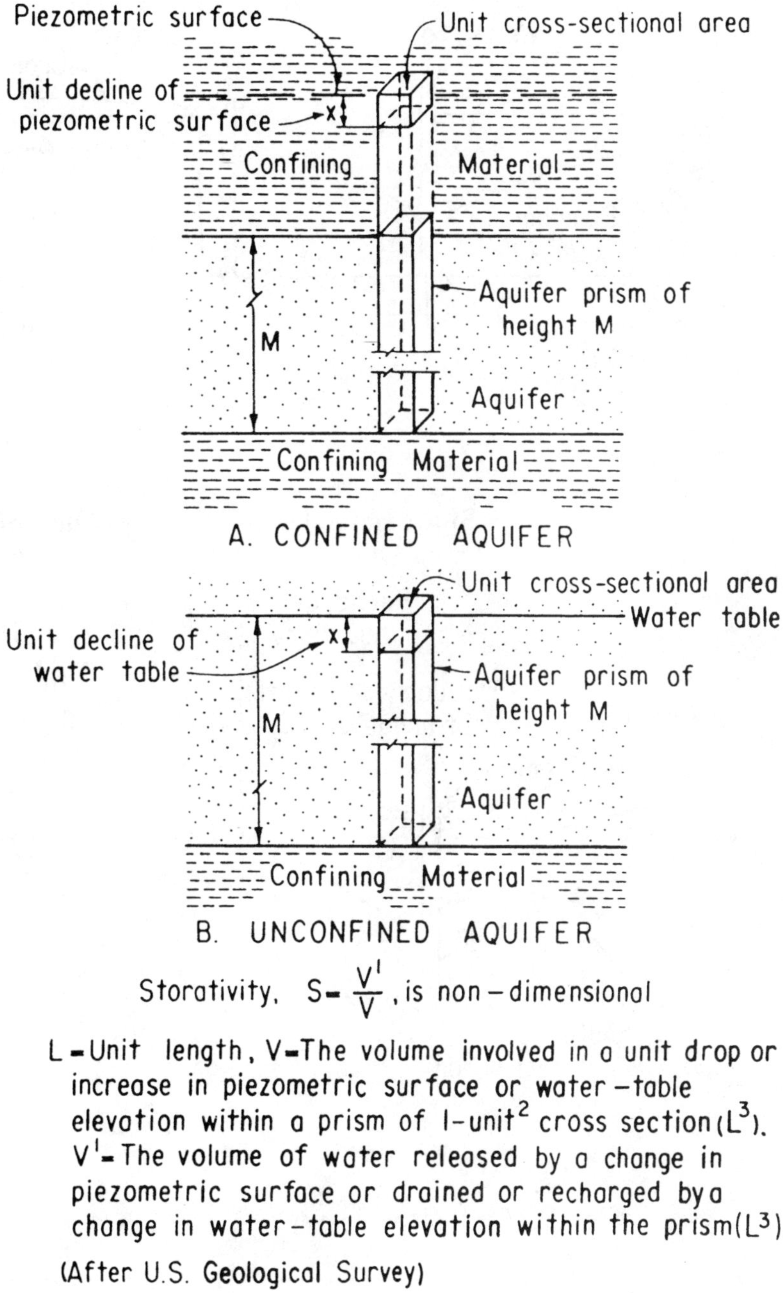

Illustration of storativity.

FIGURE A-4 Illustration of storativity.

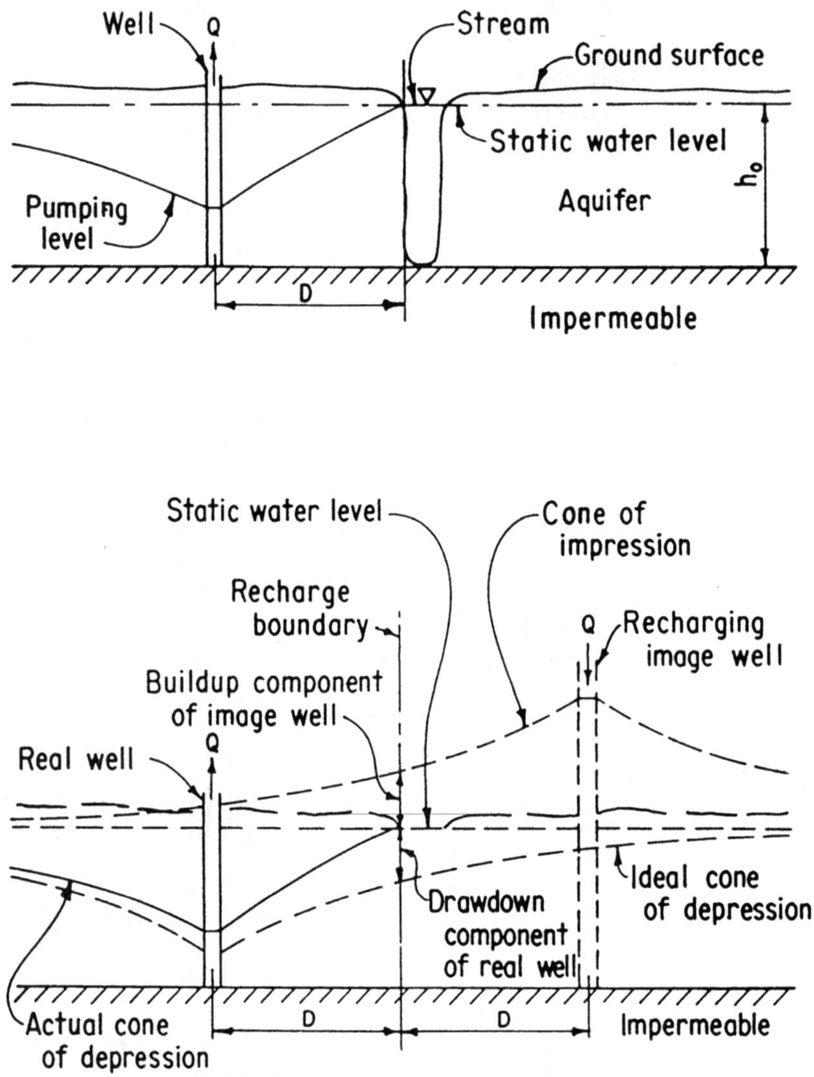

(after U.S. Geological Survey)

Relationship of an impermeable boundary and an image well.

FIGURE A-5 Relationship of an impermeable boundary and an image well.

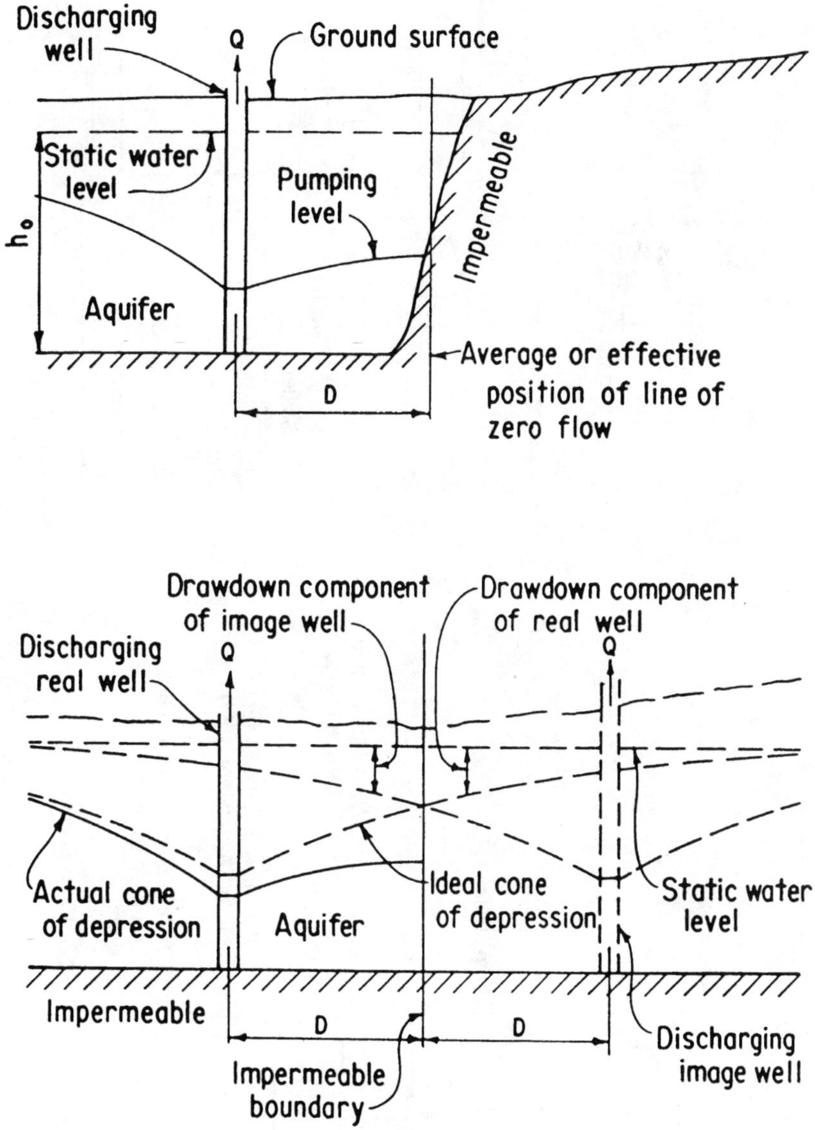

(After U.S. Geological Survey)

Relationship of a recharge boundary and an image well.

FIGURE A-6 Relationship of a recharge boundary and an image well.

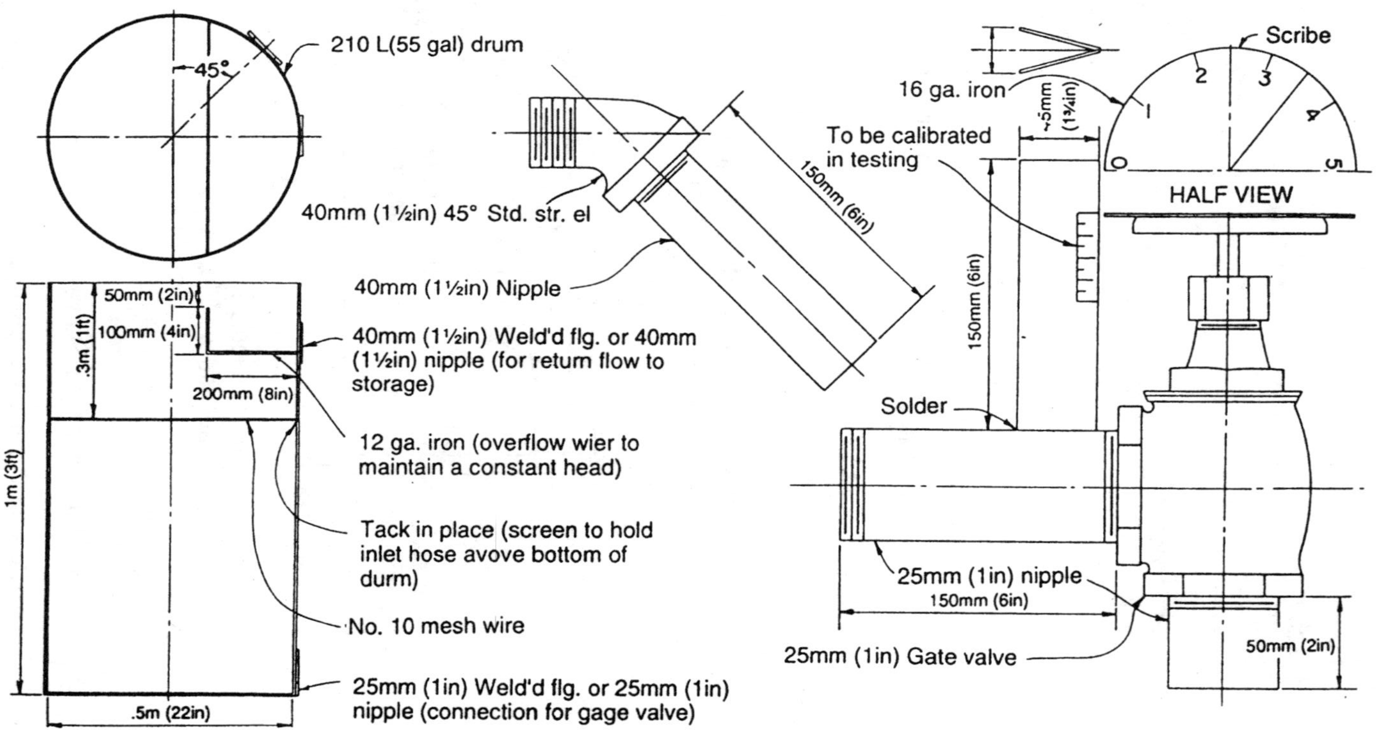

Supply tank for constant head permeability tests.

FIGURE A-7 Supply tank for constant head permeability tests.

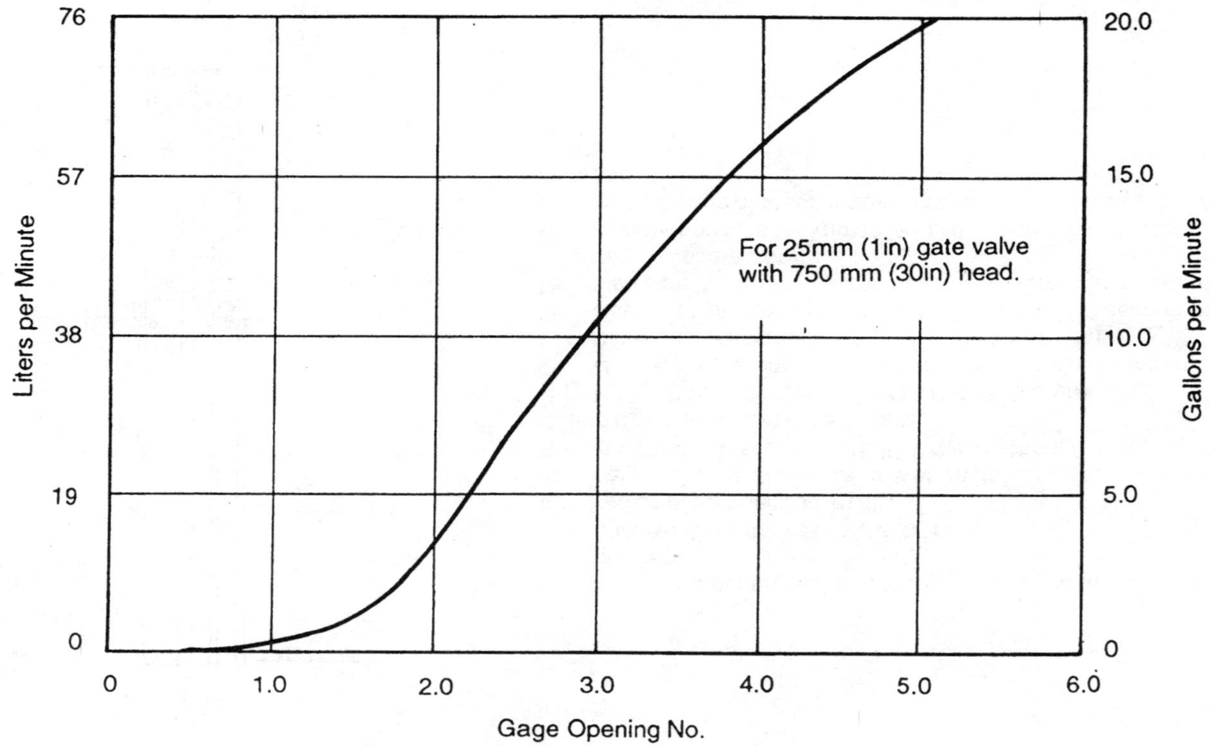

FIGURE A-8 Discharge curve for constant head supply tank.

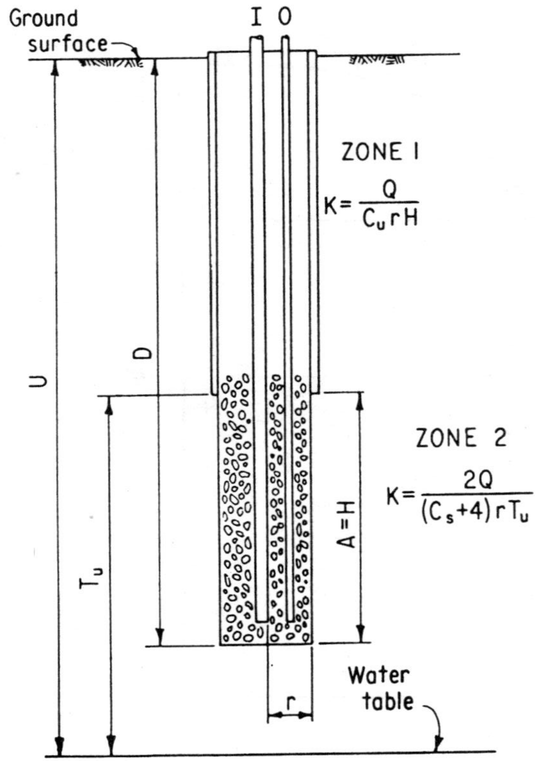

K = coefficient of permeability, meter per second (ft./sec.) under a unit gradient
Q = uniform flow into well, m^3/s ($ft^3.s$)
r = radius of test section, m (ft.)
H = height of column of water in well, m (ft.)
A = length of test section m (ft.)(for this method, A = H)
C_u and C_s = conductivity coefficients
X = H/T_u (100) = percent of unsaturated stratum
T_u = U - D + H = distance from water surface in well to water table, m (ft.)
U = thickness of unsaturated permeable bed, m (ft.)
D = distance from ground surface to bottom of test section, m (ft.)
I = feed pipe for pouring water into well (a 50mm (2 in.) standard pipe is usually satisfactory)
O = observation pipe (30mm (1¼ in.) o.d. pipe is satisfactory)
a = surface area of test section (area of wall plus area of bottom), m (ft.)
Limitations:
$A \geq 10r$ and $Q/a \geq 0.10$

Gravity permeability test (method 1) (Zangar, 1953).

FIGURE A-9 Gravity permeability test (method 1). *(Zangar, 1953)*

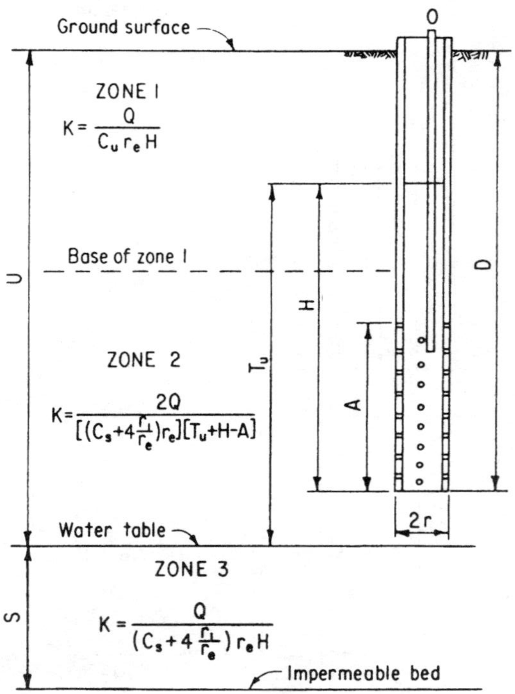

K = coefficient of permeability, m (ft.) per second under a unit gradient
Q = steady flow into well, m^3/s ($ft.^3/s$)
H = height of water in well, m (ft.)
A = length of perforated section, m (ft.)
r_1 = outside radius of casing (radius of hole in consolidated material), m (ft.)
r_e = effective radius of well = r, (area of perforations)/(outside area of perforated section of casing); r_1 = r_e in consolidated material that will stand open and is not cased
C_u and C_s = conductivity coefficients
T_u = distance from water level in casing to water table, m (ft.)
a = surface area of test section (area of perforations plus area of bottom), m^2 (ft^2) where clay seal is used at bottom, a = area of perforations
S = thickness of saturated permeable material above an underlying relatively impermeable stratum, m (ft.)
X = $H/T_u(100)$ + percent of unsaturated stratum
U = thickness of unsaturated material above water table, m (ft.)
D = distance from ground surface to bottom of test section, m (ft.)
O = observation pipe, 25 to 32mm (1 to 1¼ in.) pipe

Limitations:

$S \geq 5A$, $A \geq 10r$, and $Q/a \geq 0.10$

Notes:

In zone 3, H is the difference in elevation between the normal water table and the water level in the well. In zones 2 and 3, if a clay seal is placed at the bottom of the casing, the factor $4(r_1/r_e)$ is omitted from the equations. Where the test is run with "A" as an open hole, $r_1/r_e = 1$ and $(C_s+4(r_1/r_e)) = (C_s+r)$.

Gravity permeability test (method 2).

FIGURE A-10 Gravity permeability test (method 2).

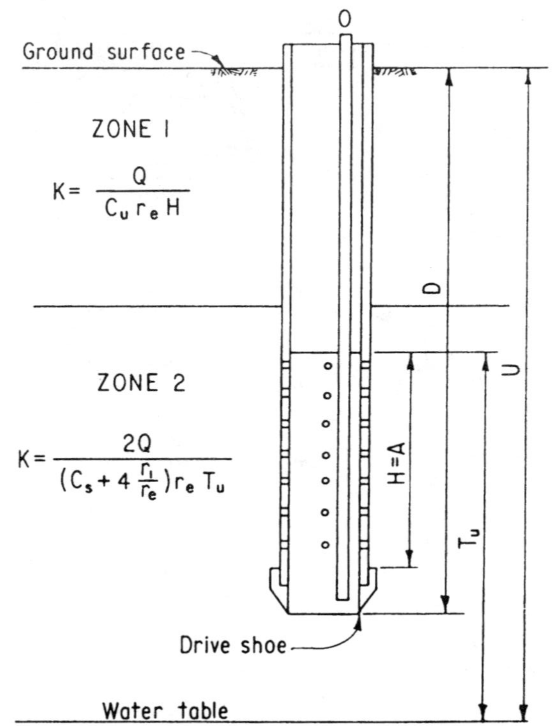

k = coefficient of permeability, m (ft) per second under a unit gradient
Q = steady flow into well m^3/s (ft^3/s)
r_1 = outside radius of casing
r_e = effective radius of casing = r_1 (area of perforations)/(outside area of A)
A = length of perforated section, m (ft)
C_u and C_s = conductivity coefficients
H = height of column of water in perforated section, m (ft)
T_u = distance from water level in casing to water table, m (ft)
X = $H/T_u(100)$ = percent of unsaturated stratum
O = observation pipe(30mm (1¼in) o.d. pipe is satisfactory)
U = thickness of unsaturated material above water table, m (ft)
D = distance from ground surface to bottom of test section, m (ft)
a = surface area of test section (area of perforations plus area of bottom), m^2 (ft^2); where clay seal is used at bottom, a = area of perforations

Limitations:
$Q/a \leq 0.10$, $A \geq 10r$

Note:
In zone 2, if clay seal is placed at bottom of casing, the factor $(4)r_1/r_e$ is omitted from equation

Gravity permeability test (method 3) (Zangar, 1953).

FIGURE A-11 Gravity permeability test (method 3). *(Zangar, 1953)*

HOLE NO. E - 4 **LOCATION** SAMPLE FARM

OBSERVER D.M.S. **DATE** AUGUST 6, 1992

HOLE: CASED [X] **UNCASED** [ ]

HOLE DIAMETER 102 **MILLIMETERS** (4 inches)

LOG **DESCRIPTION**

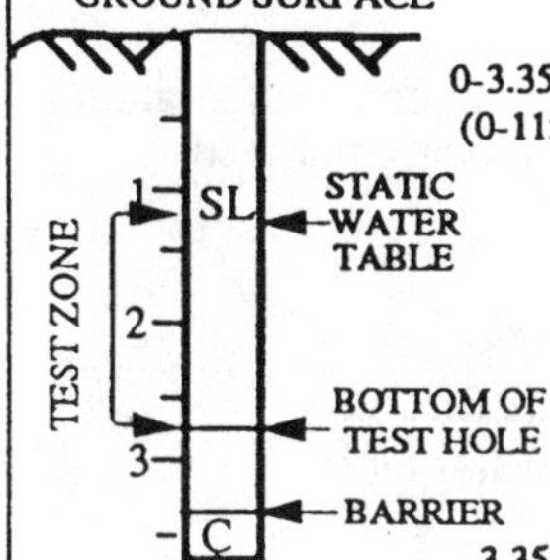

0-3.35 m: (0-11ft) Light brown sandy loam(SL), friable, nonsticky, granular. Wet at 1.52 m. (5ft) Slightly compacted below 1.83 m (6ft). Appears to have good hydraulic conductivity.

3.35-3.66 m: (11-12ft) Blue gray clay (C), Sticky, structureless. Appears to be impermeable.

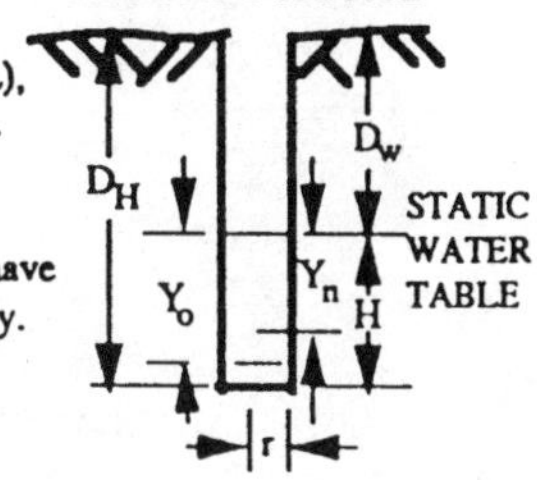

$r = 0.051$ m (0.167 ft)

$D_H = 2.74$ m (9.0 ft)

$D_w = 1.46$ m (4.8 ft)

$H = 1.28$ m (4.2 ft)

$Y_o = 0.96$ m (3.15 ft)

$0.8\,Y_o = 0.77$ m (2.52 ft)

	TIME SECONDS	Δt	Y_n meters	(feet)	ΔY meters	(feet)
	0	--	0		--	
	13	13	0.960	(3.15)	Y_o	
	23	10	0.927	(3.04)	0.033	(0.11)
	33	10	0.893	(2.93)	0.034	(0.11)
	43	10	0.860	(2.82)	0.033	(0.11)
	53	10	0.823	(2.70)	0.037	(0.12)
	63	10	0.789	(2.59)	0.034	(0.11)
$0.8\,Y_o$	73	10	0.759	(2.49)	0.030	(0.10)
	83	10	0.732	(2.40)	0.027	(0.09)
	93	10	0.704	(2.31)	0.028	(0.09)

$\bar{Y}_n = \frac{0.960 + 0.759}{2} = 0.860$ meter (2.82 feet) $\qquad \frac{\bar{Y}_n}{r} = \frac{0.860}{0.051} = 16.86$

$\Delta Y = 0.0335$ meter (0.11 ft) $\qquad C = 390$ (from chart)

$\Delta t = 10$ seconds $\qquad K = C\,\frac{\Delta Y}{\Delta t} = 1.31$ meters (4.3 feet) per day

$\frac{H}{r} = \frac{1.28}{0.051} = 25.10$ $\qquad$ or 5.45 centimeters (2.151 inches) per hour

-Data and computation sheet on auger-hole test for hydraulic conductivity.

FIGURE A-12 Data and computation sheet on auger-hole test for hydraulic conductivity.

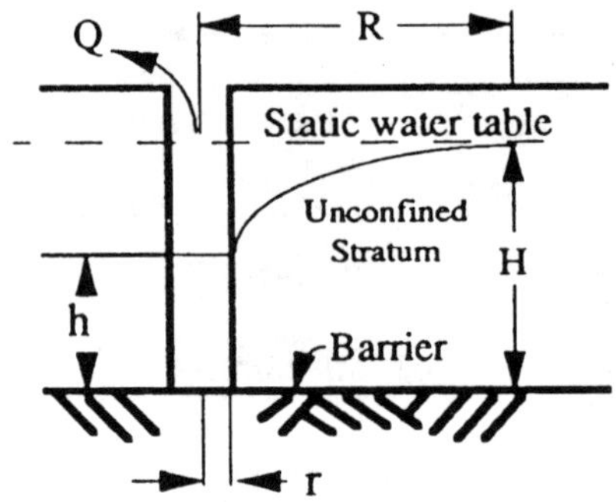

$$K = \frac{Q \log_e (R/r)}{\pi (H^2 - h^2)}$$

Assume R = 500 x r
for most cases.

(a) Pumping from a uniform unconfined stratum, water table in stratum being pumped.

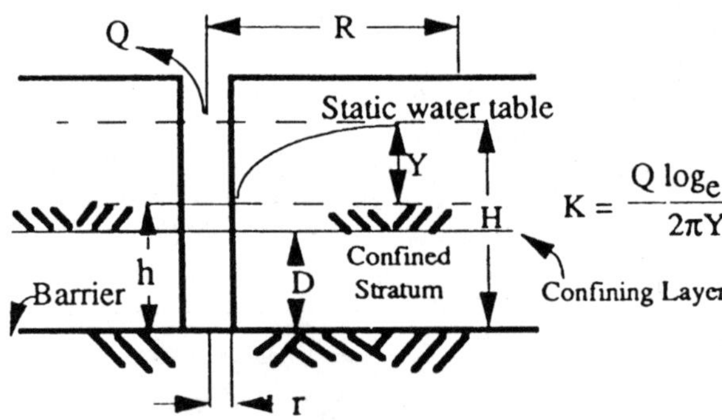

$$K = \frac{Q \log_e (R/r)}{2\pi YD}$$

(b) Pumping from a confined stratum, water table above stratum being pumped.

K = Hydraulic conductivity, m^3/m^2/day (ft^3/ft^2/day)
Q = Flow rate at steady state conditions, m^3/day (ft^3/day)
Y = Drawdown from static water surface = H-h, m (ft)
H = Height of static water table above bottom of hole, m (ft)
h = Depth of water in hole at steady state pumping conditions, m (ft)
D= Flow thickness of strata between bottom of the hole and overlying (confining) stratum, m (ft)
R = Distance from centerline of well to point of zero drawdowm, m (ft)
r = Effective radius of well, m (ft)

Determination of hydraulic conductivity by pumping from a uniform or confined stratum.

FIGURE A-13 Determination of hydraulic conductivity by pumping from a uniform or confined stratum.

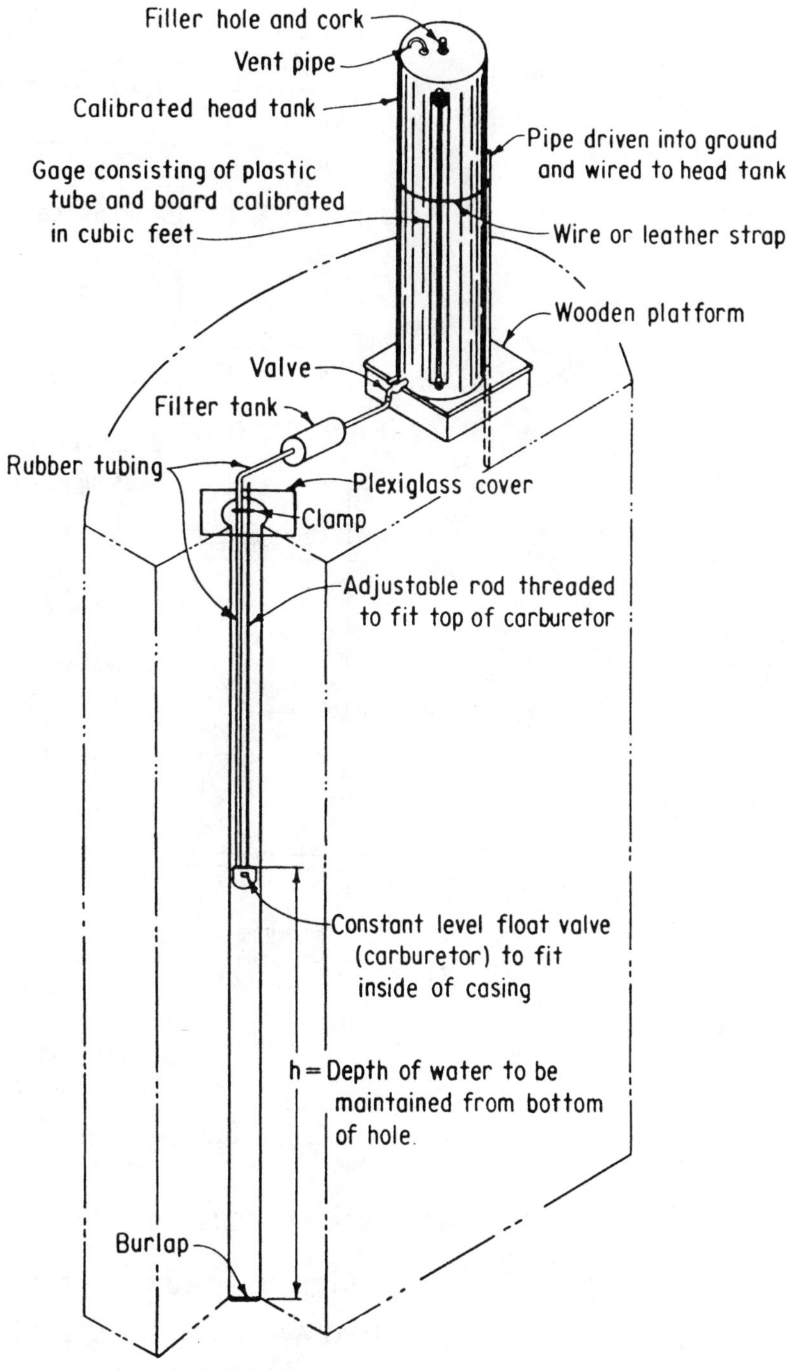

–Equipment setup for a shallow well pump-in test.

FIGURE A-14 Equipment setup for a shallow well pump-in test.

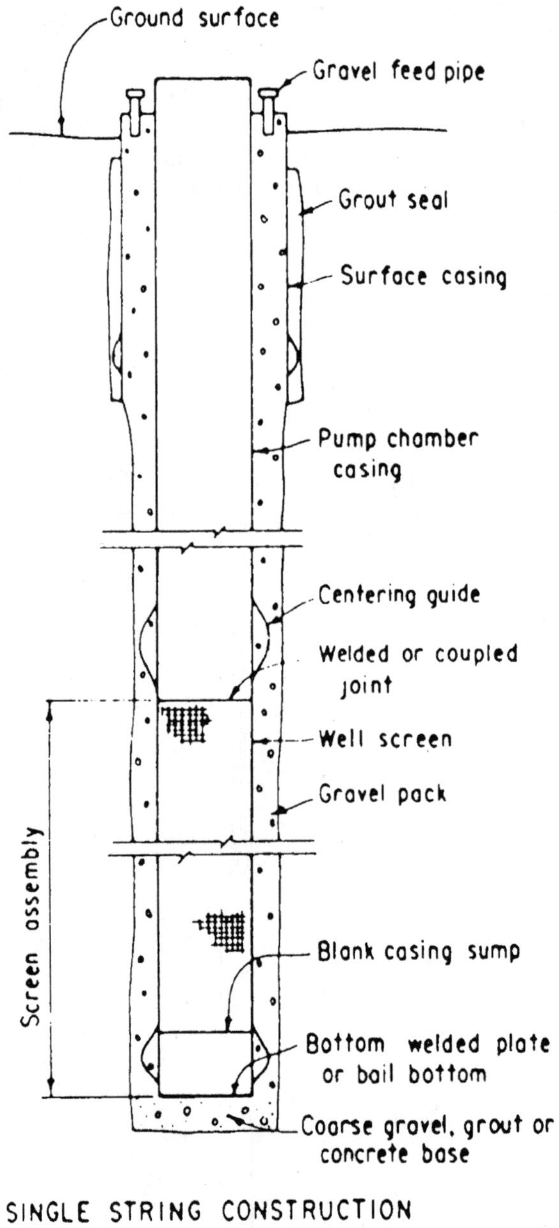

Gravel-packed, rotary drilled well for single string construction.

FIGURE A-15a Gravel-packed, rotary drilled well for single string construction.

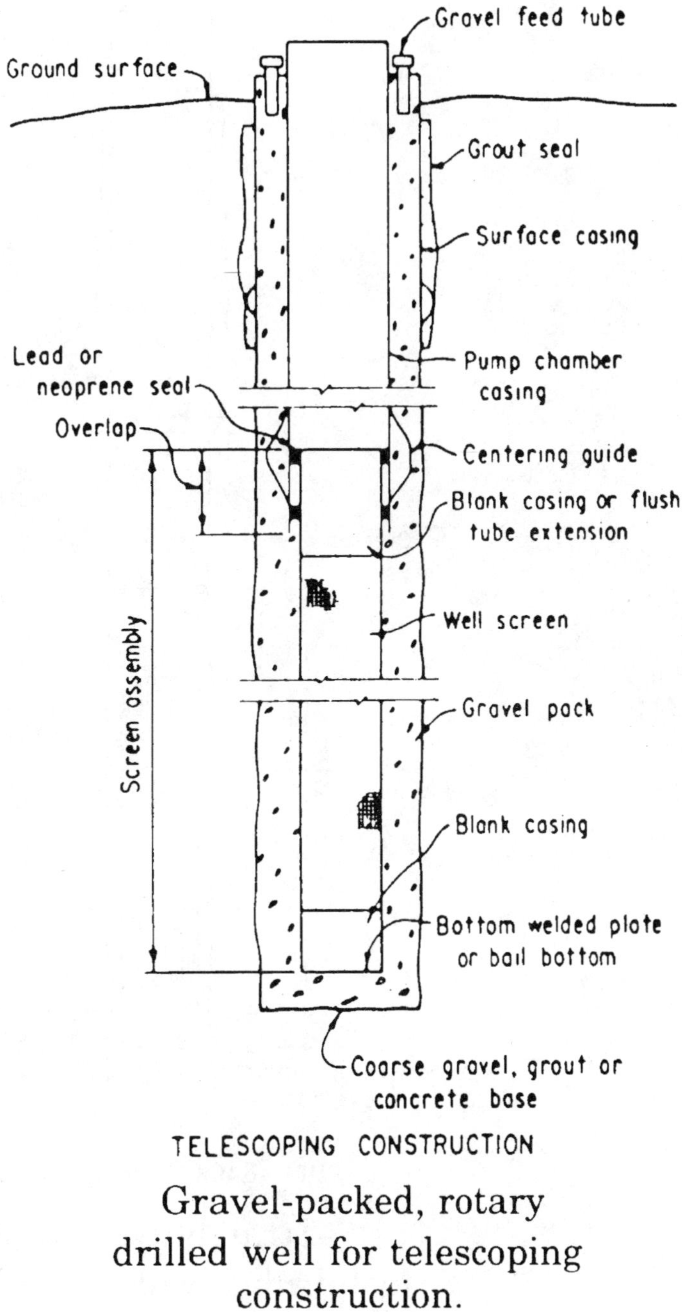

FIGURE A-15b Gravel-packed, rotary drilled well for telescoping construction.

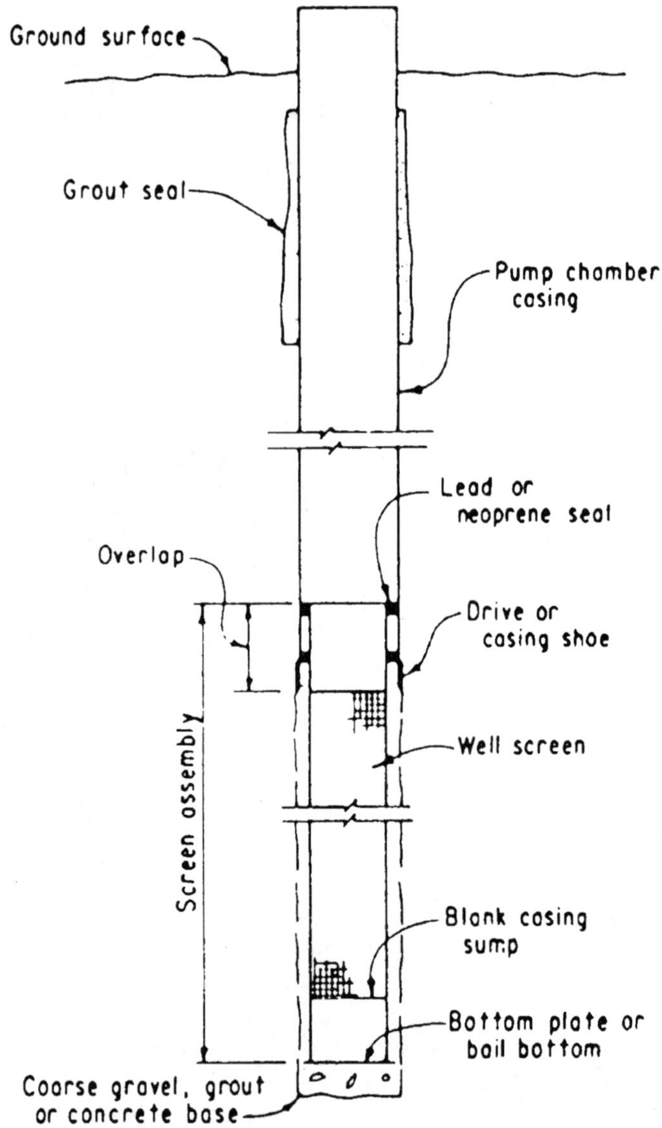

PULL BACK CONSTRUCTION

-Straight wall, cable tool drilled well for pull back construction.

FIGURE A-15c Straight wall, cable tool drilled well for pull back construction.

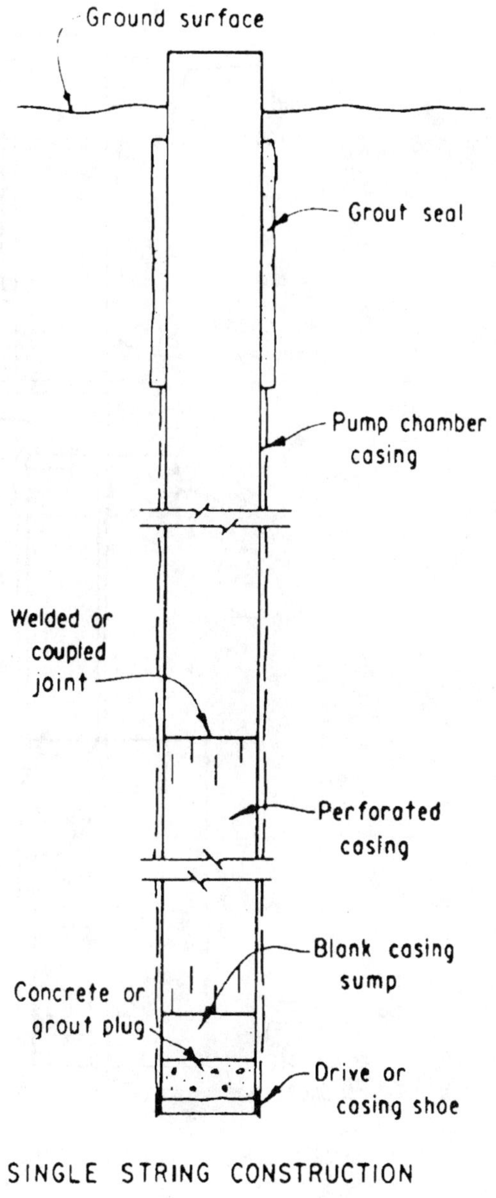

Straight wall, cable tool drilled well for single string construction.

FIGURE A-16a Straight wall, cable tool drilled well for single string construction.

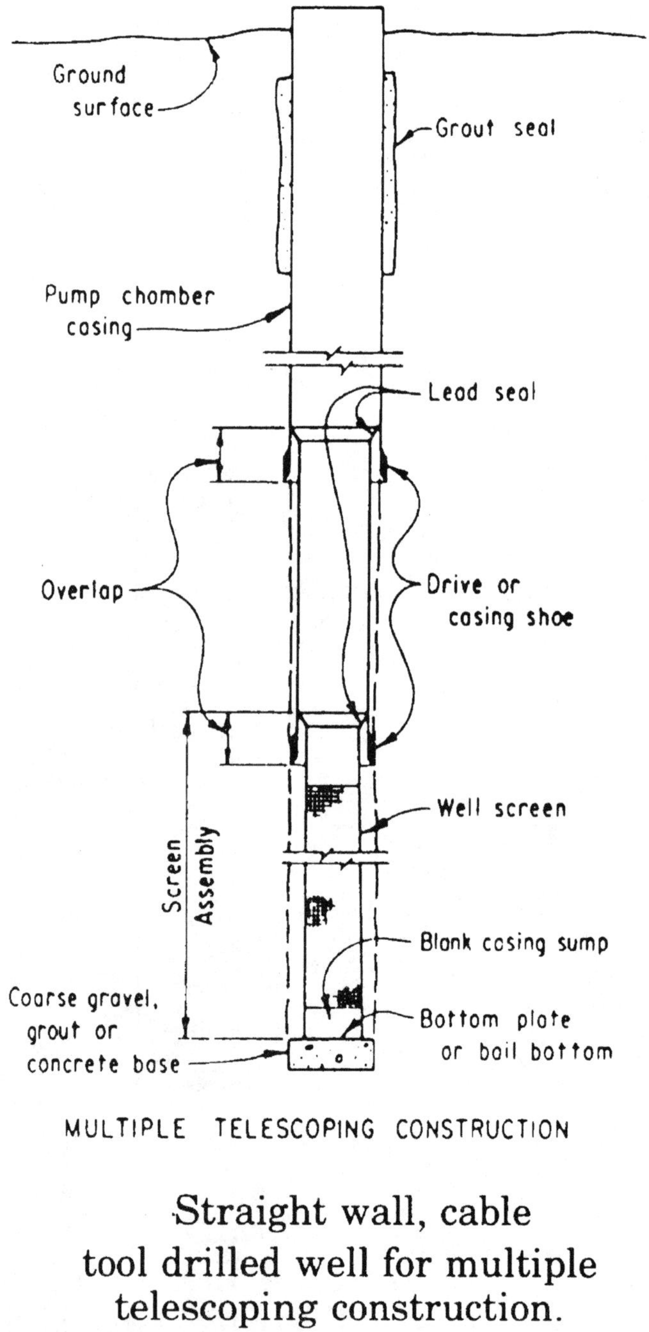

FIGURE A-16b Straight wall, cable tool drilled well for multiple telescoping construction.

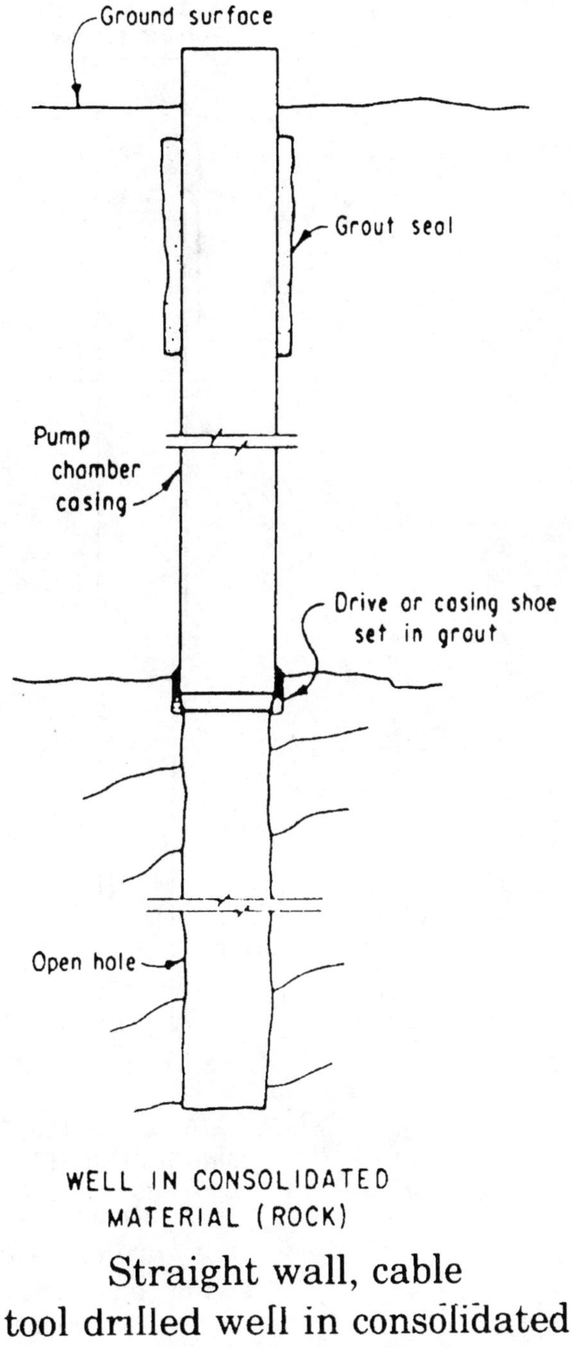

FIGURE A-16c Straight wall, cable tool drilled well in consolidated material (rock).

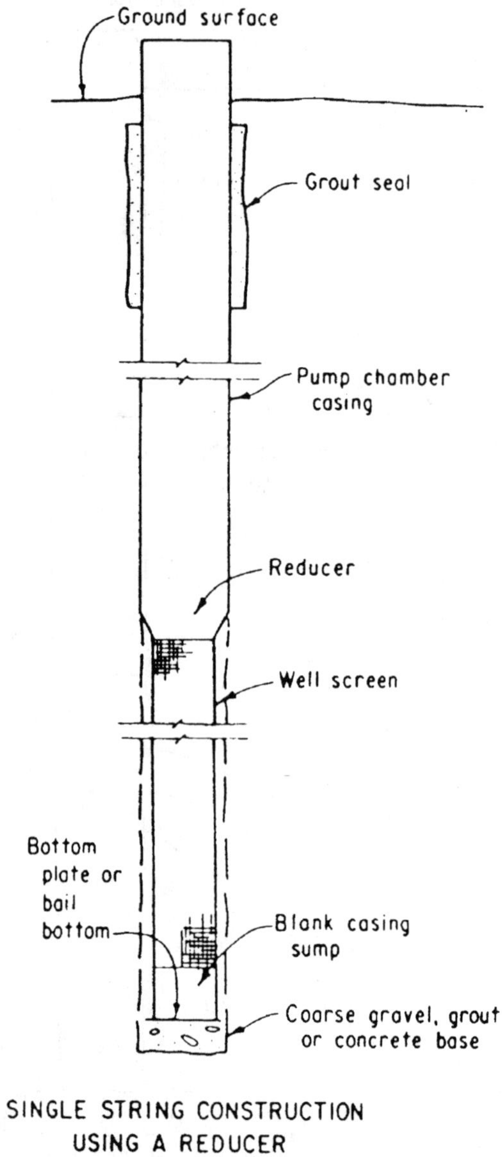

Straight wall, rotary drilled well for single string construction using a reducer.

FIGURE A-17a Straight wall, rotary drilled well for single string construction using a reducer.

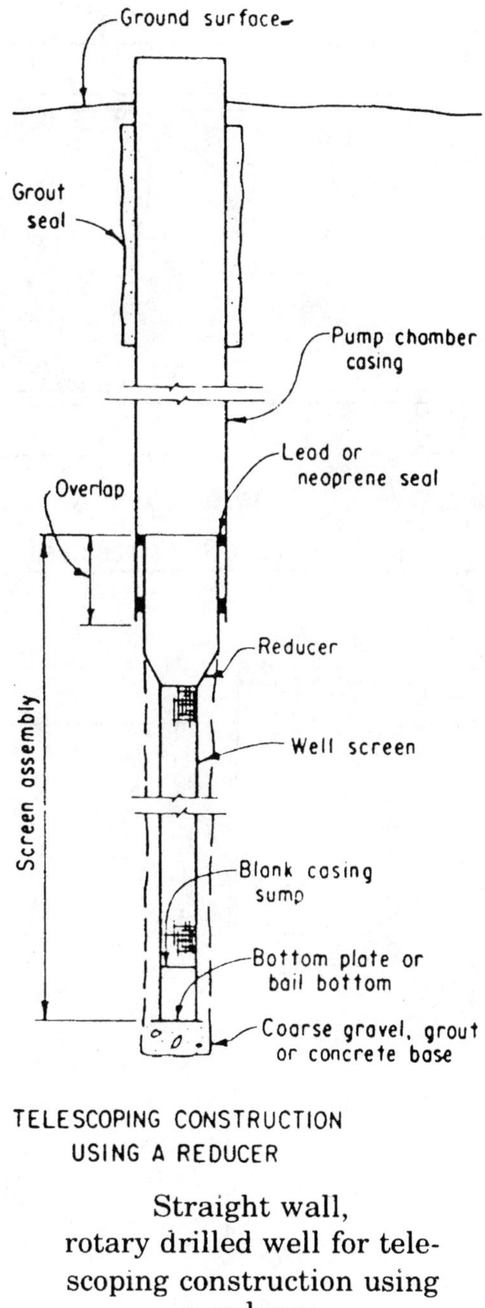

FIGURE A-17b Straight wall, rotary drilled well for telescoping construction using a reducer.

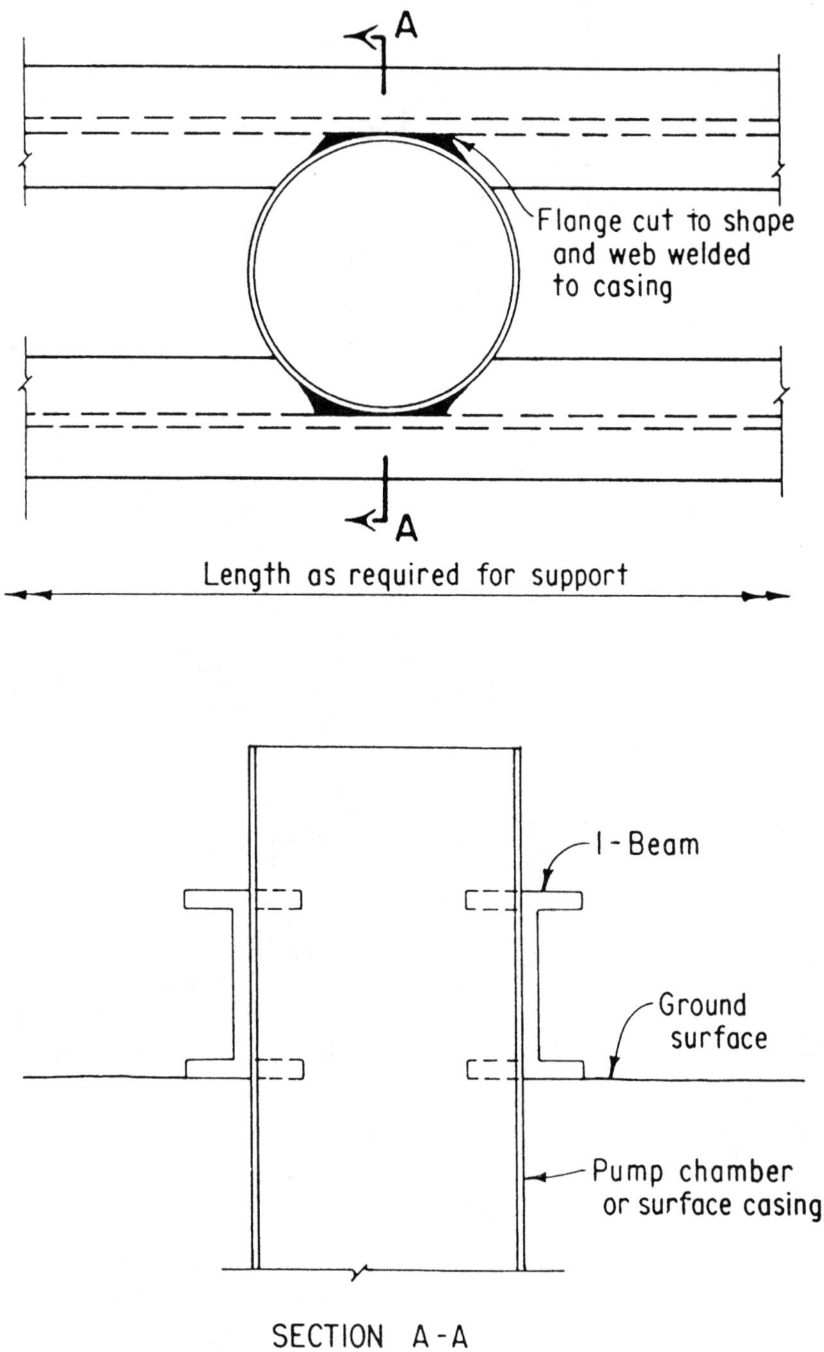

-Support of a casing by I-beams.

FIGURE A-18 Support of a casing by I-beams.

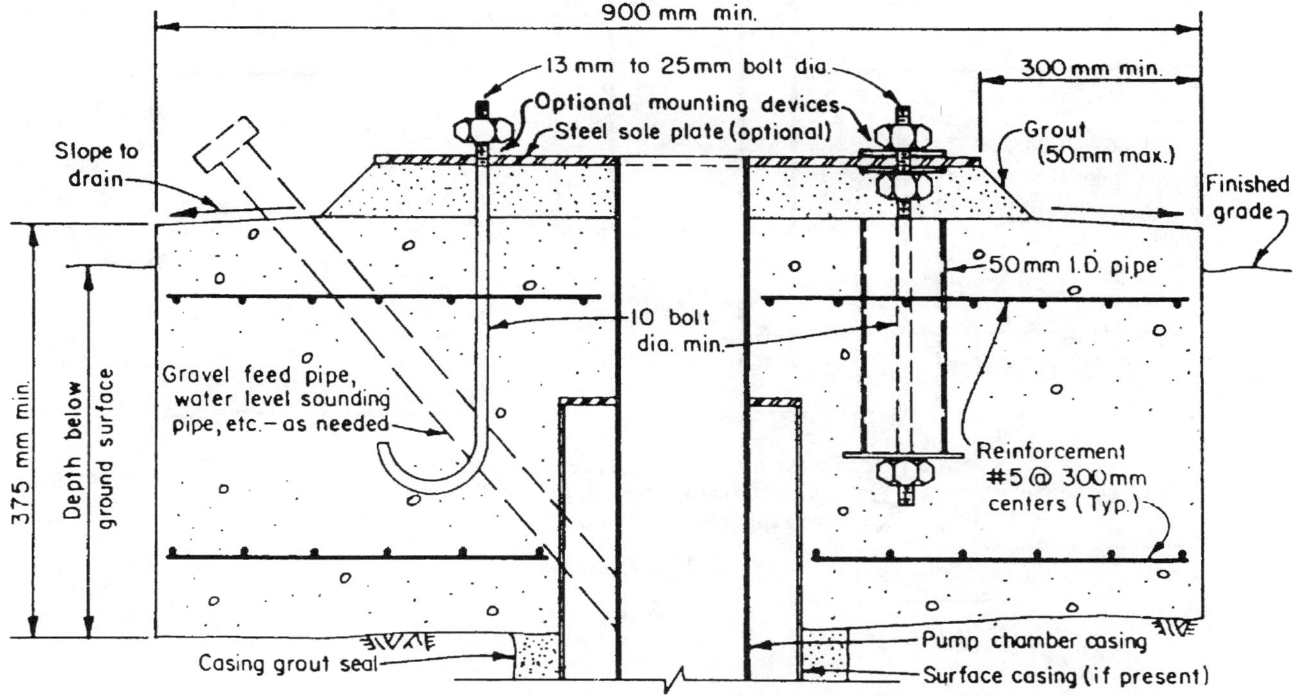

FIGURE A-19a Schematic section of a typical concrete pump foundation (millimeters).

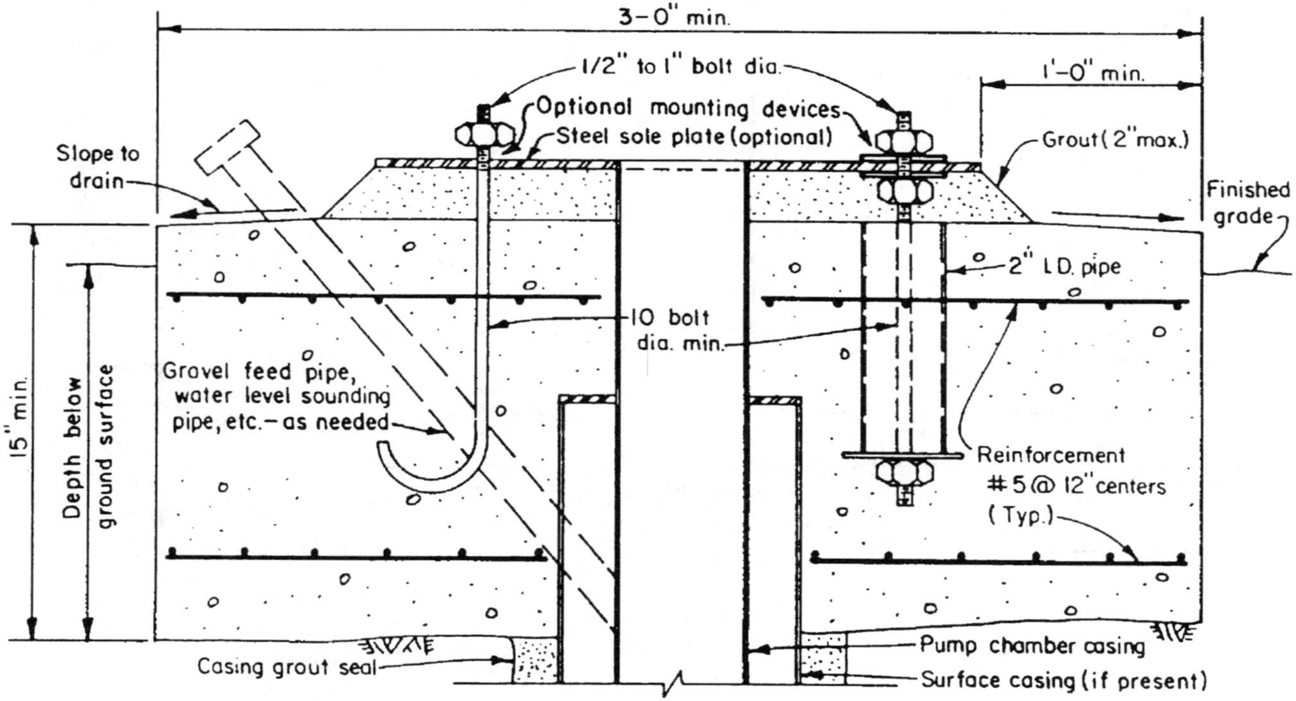

FIGURE A-19b Schematic section of a typical concrete pump foundation (inches).

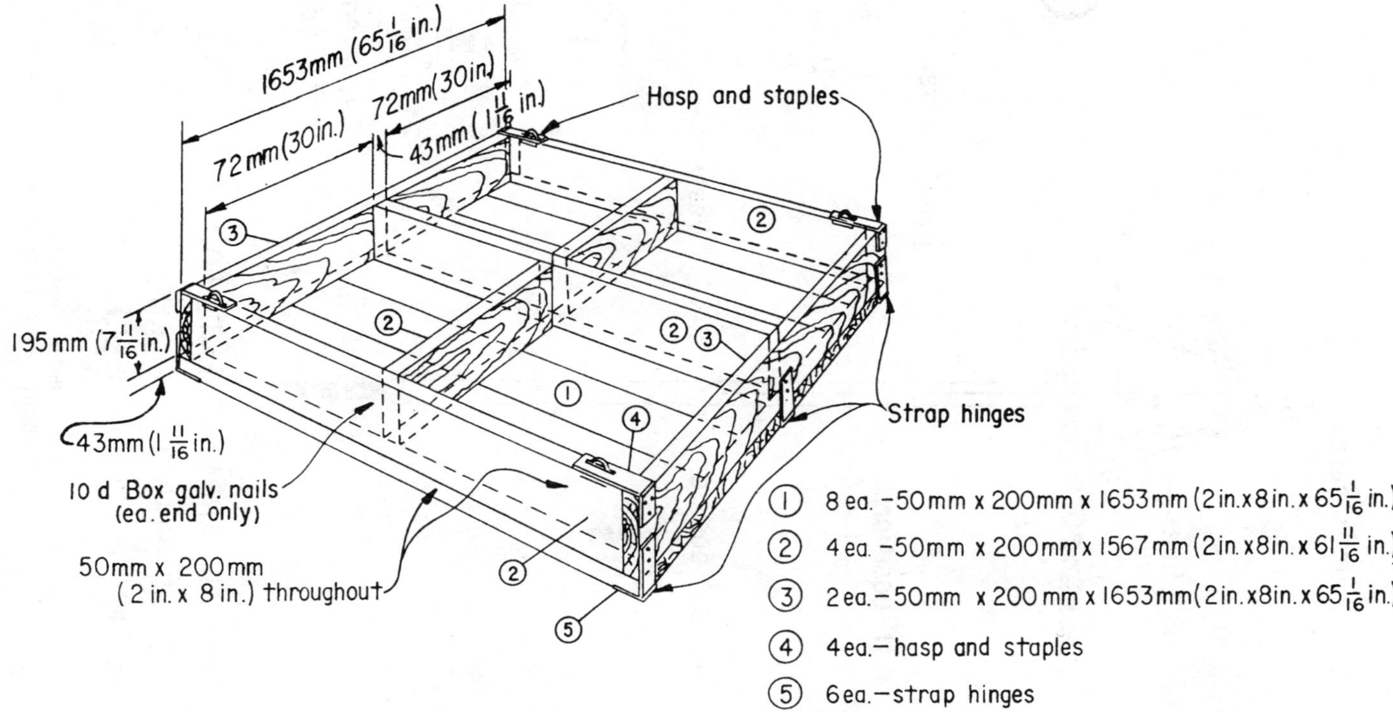

Compartmented cuttings sample box for use with a cable tool drill.

FIGURE A-20 Compartmented cuttings sample box for use with a cable tool drill.

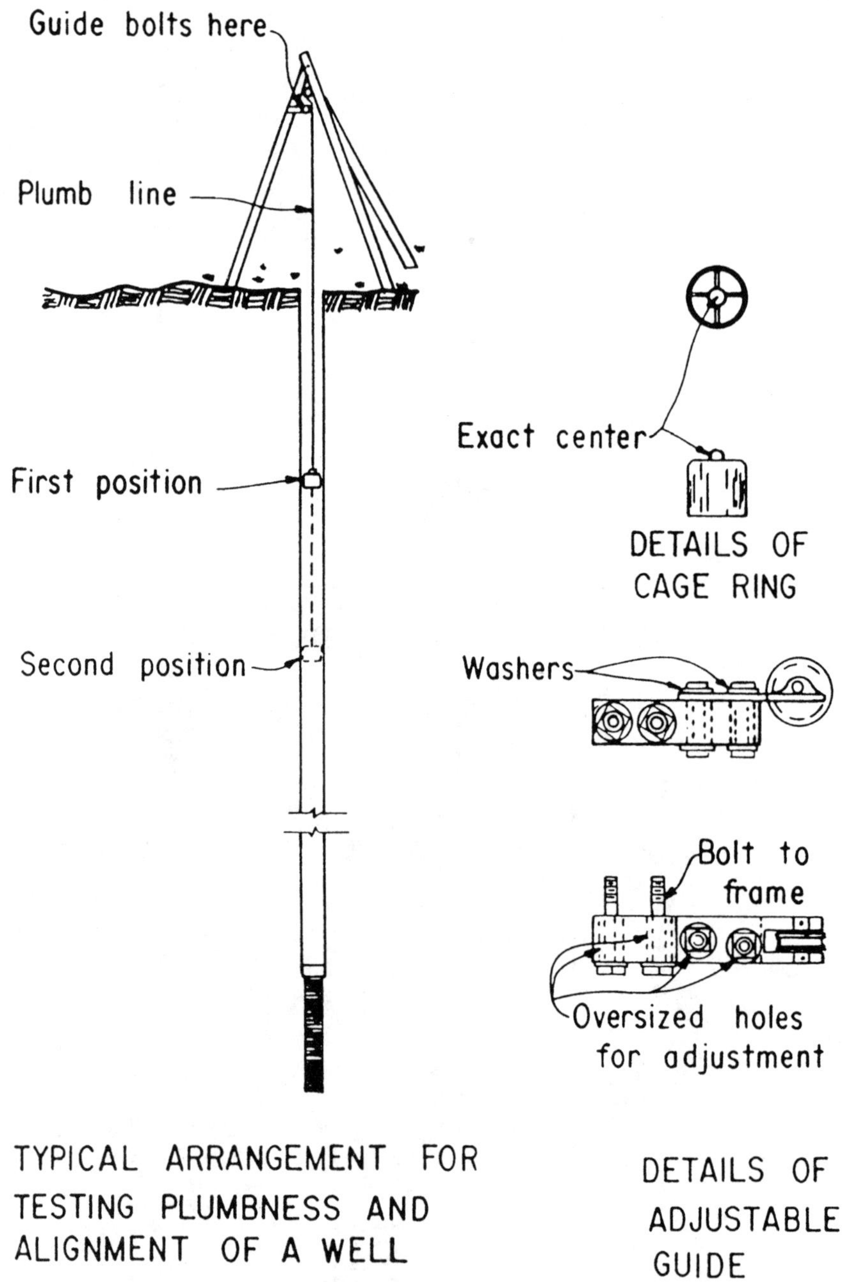

Cable suspended cage for checking straightness and plumbness of wells.

FIGURE A-21 Cable suspended cage for checking straightness and plumbness of wells.

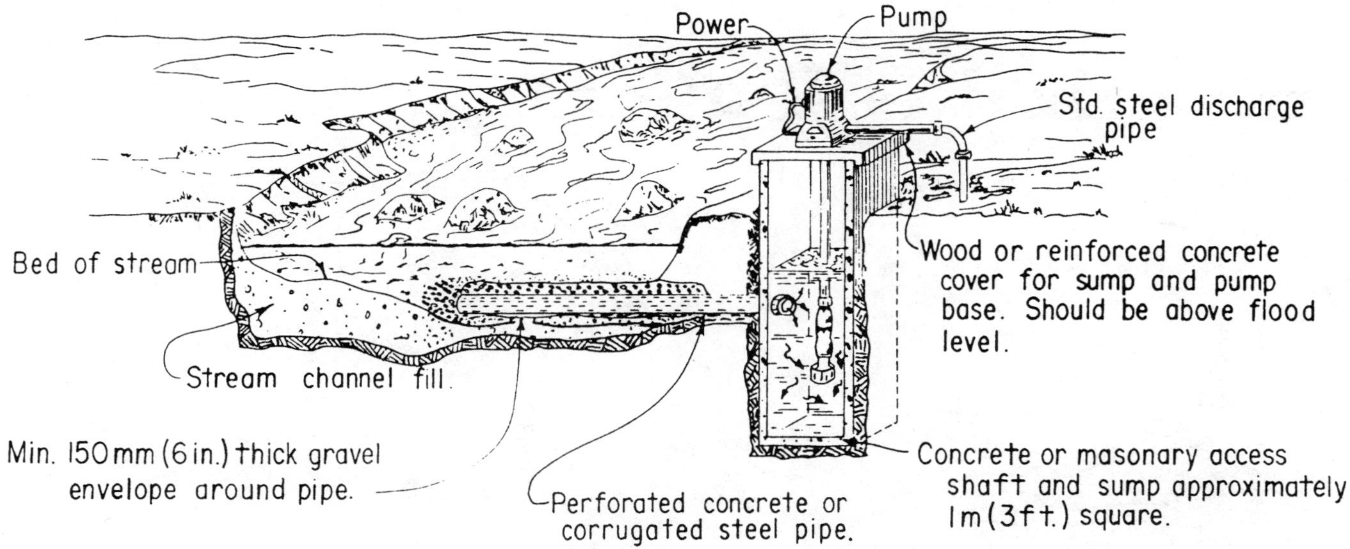

—Schematic section of an infiltration gallery constructed to intercept infiltration from a stream.

FIGURE A-22 Schematic section of an infiltration gallery constructed to intercept infiltration from a stream.

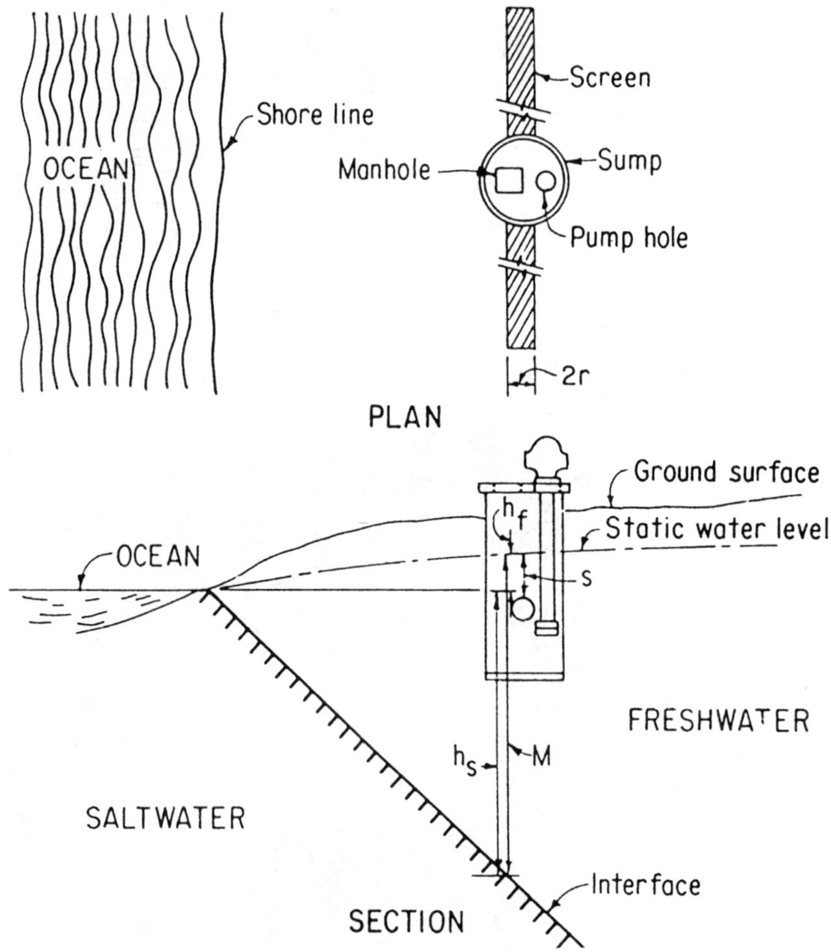

Schematic section of an infiltration gallery constructed to obtain fresh water near a seacoast.

FIGURE A-23 Schematic section of an infiltration gallery constructed to obtain fresh water near a seacoast.

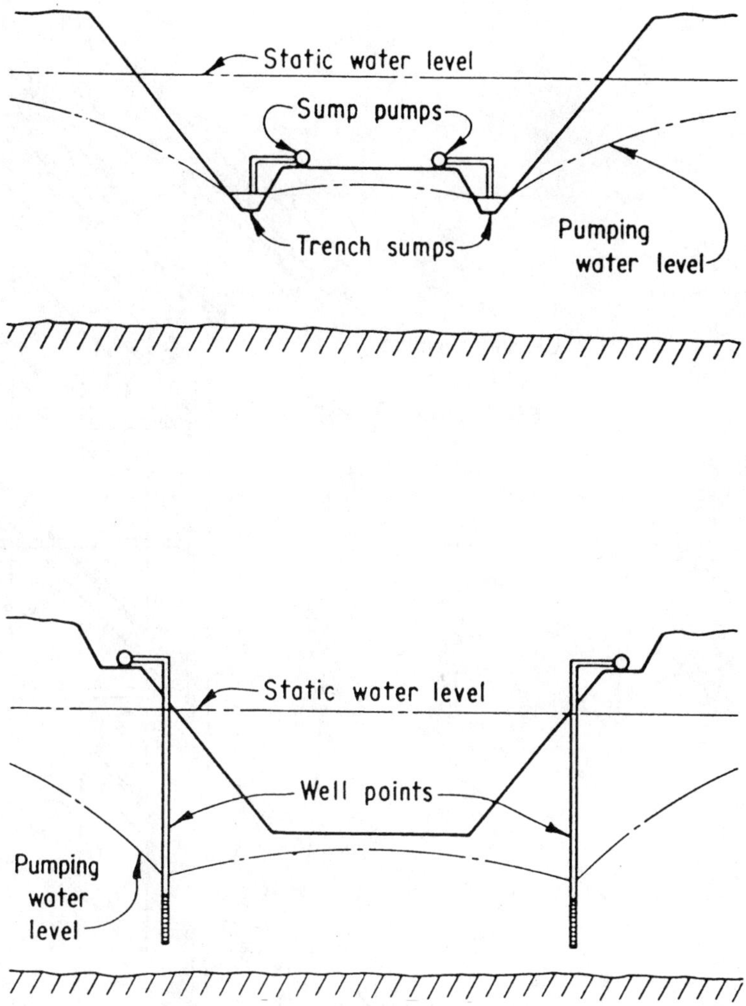

Schematic sections showing dewatering by pumping from sumps and well-point systems.

FIGURE A-24 Schematic sections showing dewatering by pumping from sumps and well-point systems.

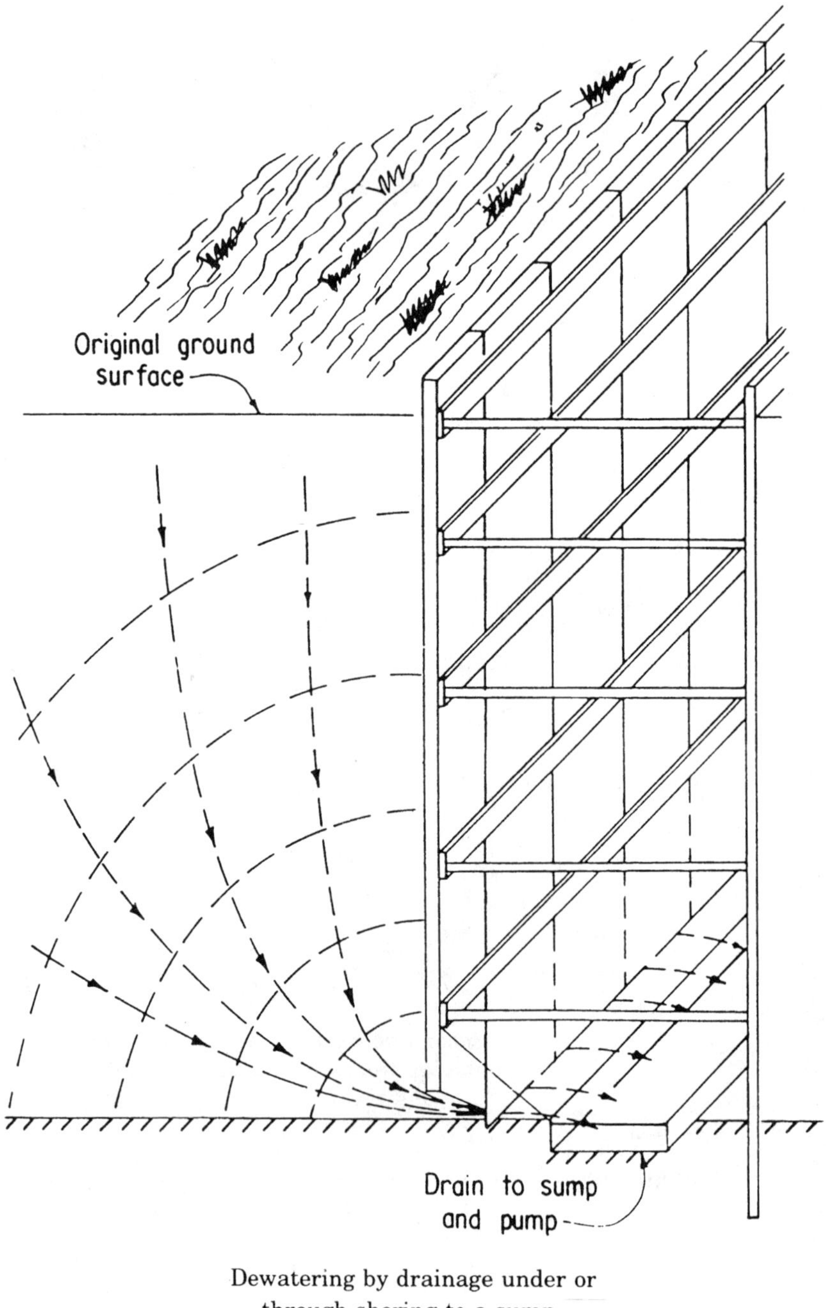

Dewatering by drainage under or through shoring to a sump.

FIGURE A-25 Dewatering by drainage under or through shoring to a sump.

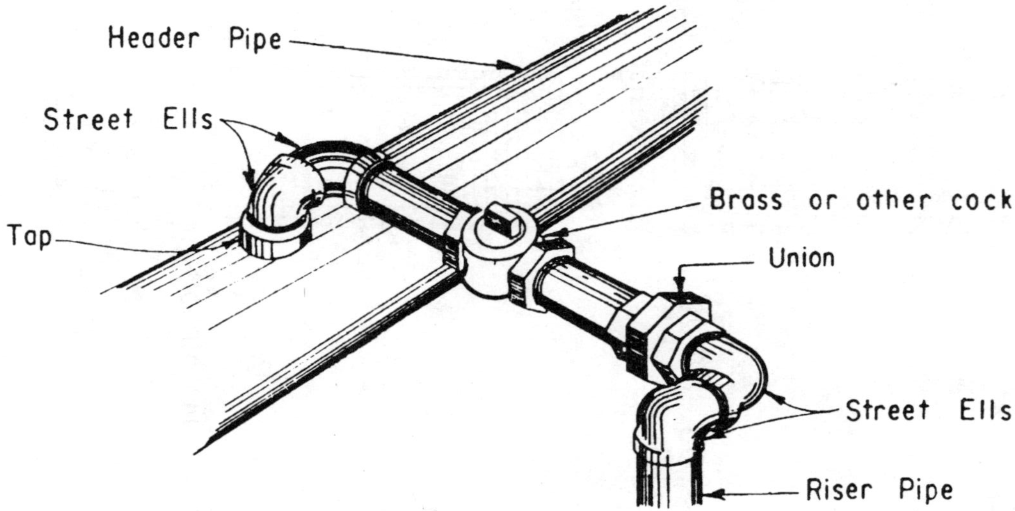

Details of connecting fitting from riser pipe to header pipe. Made up of 25 to 65 mm (1 to 2½ in.) pipe fittings, the connection may be swung horizontally and vertically to meet the top of the riser pipe. This flexibility is important since, in practice, the well point and its riser pipe generally will not match the calculated position.

Swing connection fittings for riser pipe to header pipe connection (UOP Johnson Division, 1969).

FIGURE A-26 Swing connection fittings for riser pipe to header pipe connection. *(UOP Johnson Division, 1969)*

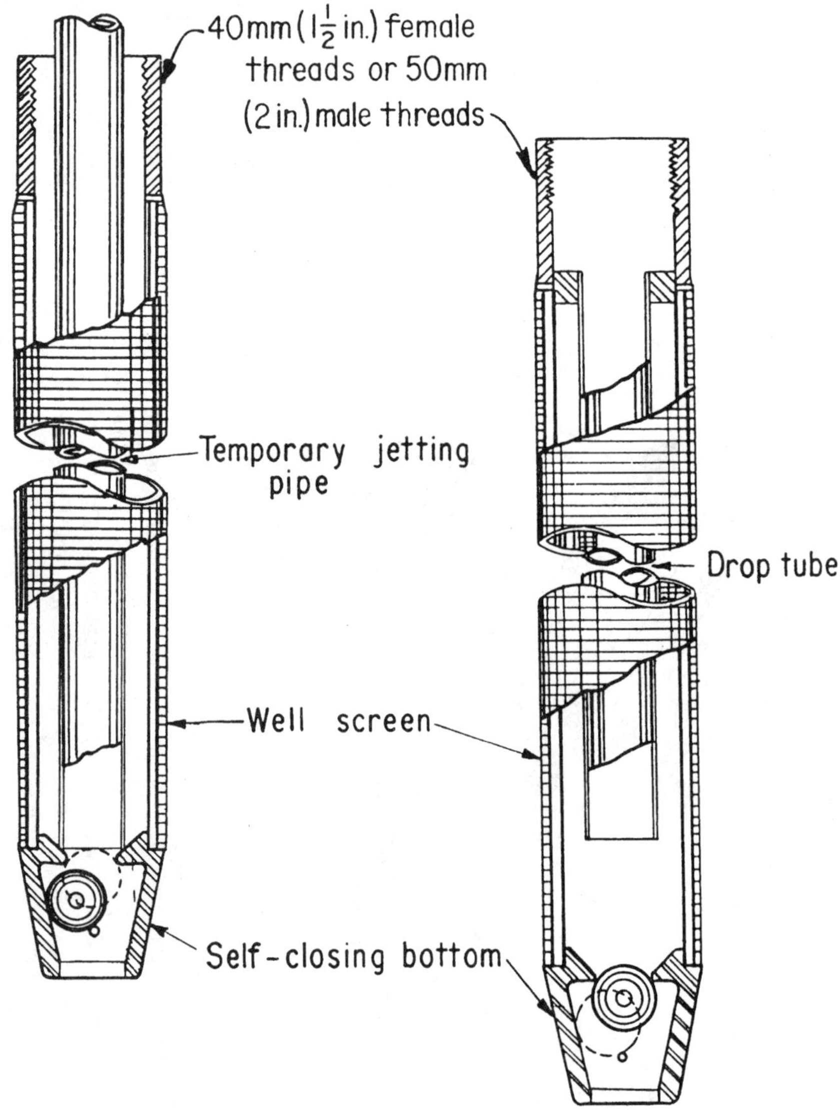

Typical well points equipped with jetting tips.
From (UOP Johnson, 1969).

FIGURE A-27 Typical well points equipped with jetting tips. *(From UOP Johnson, 1969)*

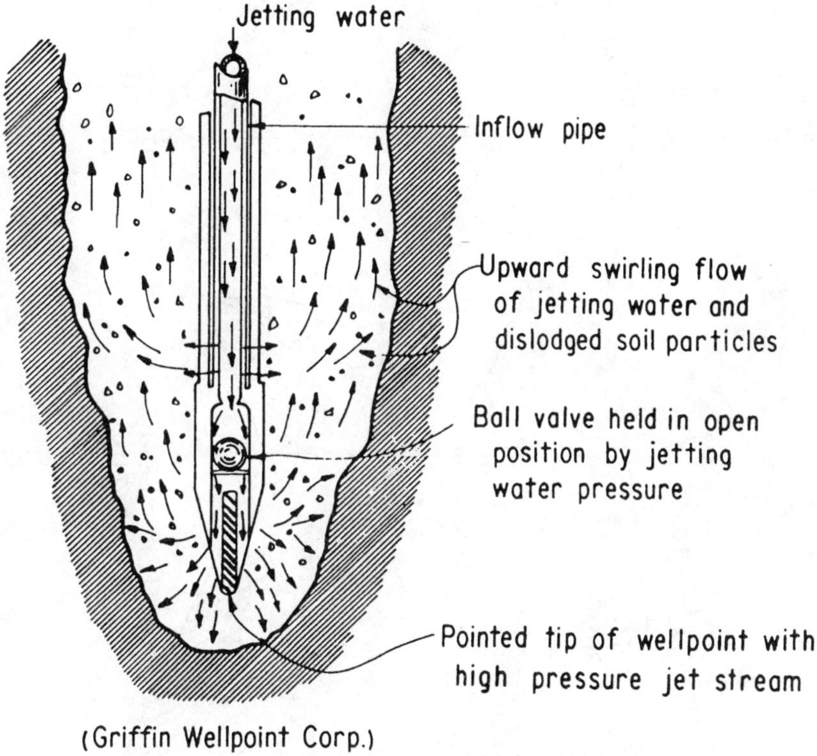

Installation of a well point by jetting (Griffing Well Point Corp.).

FIGURE A-28 Installation of a well point by jetting. *(Griffing Well Point Corp.)*

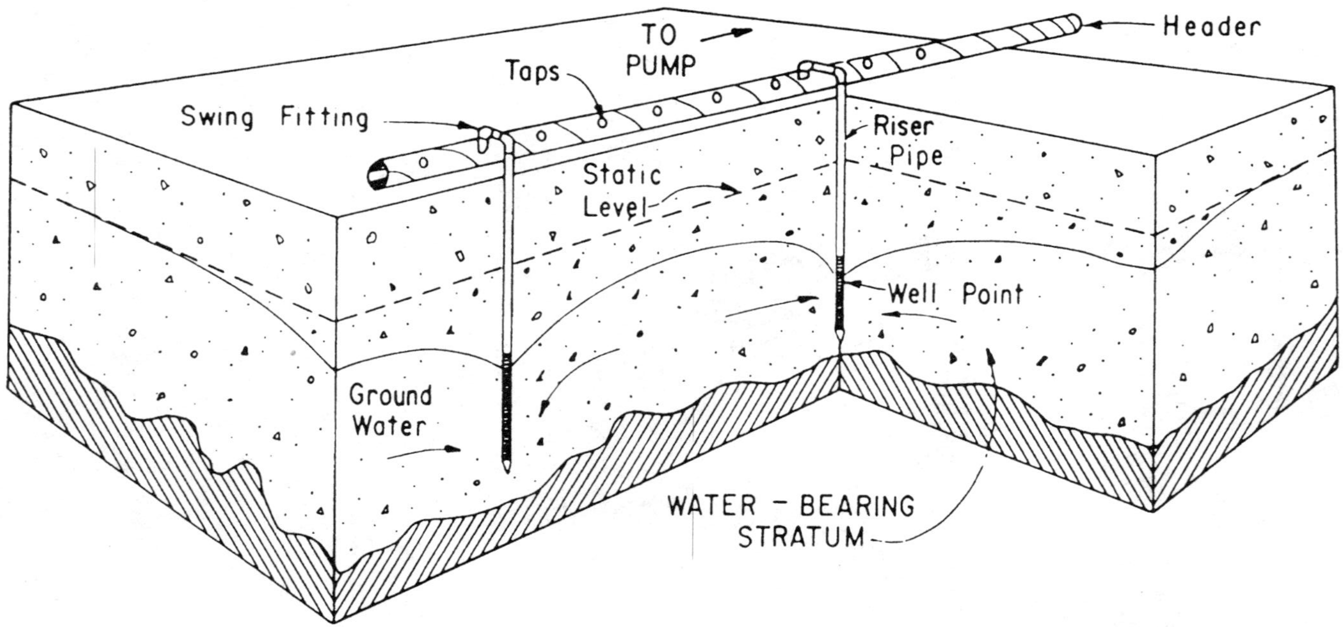

Schematic section of a portion of a well-point dewatering system (UOP Johnson Division, 1969).

FIGURE A-29 Schematic section of a portion of a well-point dewatering system. *(UOP Johnson Division, 1969)*

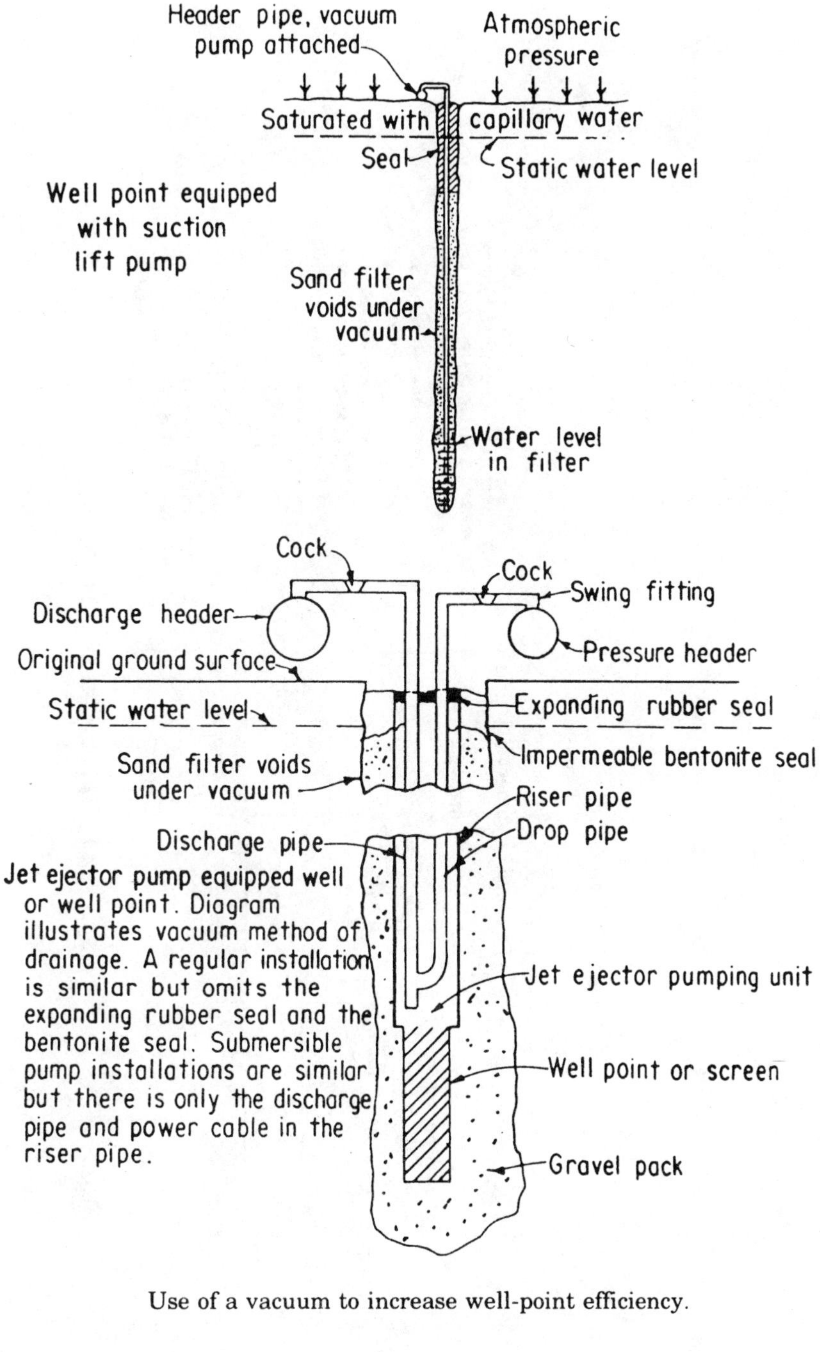

FIGURE A-30 Use of a vacuum to increase well-point efficiency.

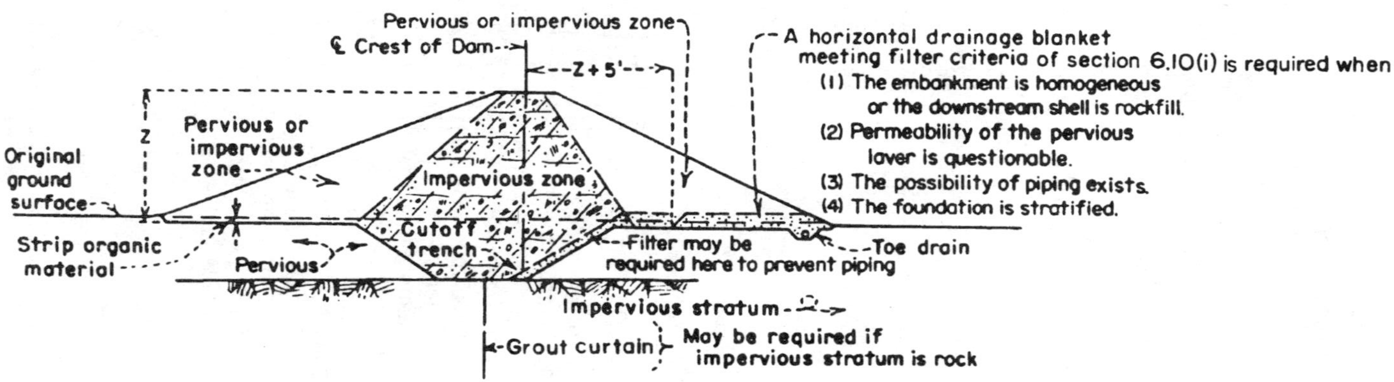

(A) SHALLOW PERVIOUS FOUNDATION

FIGURE A-31 Intermediate depth of exposed pervious foundation.

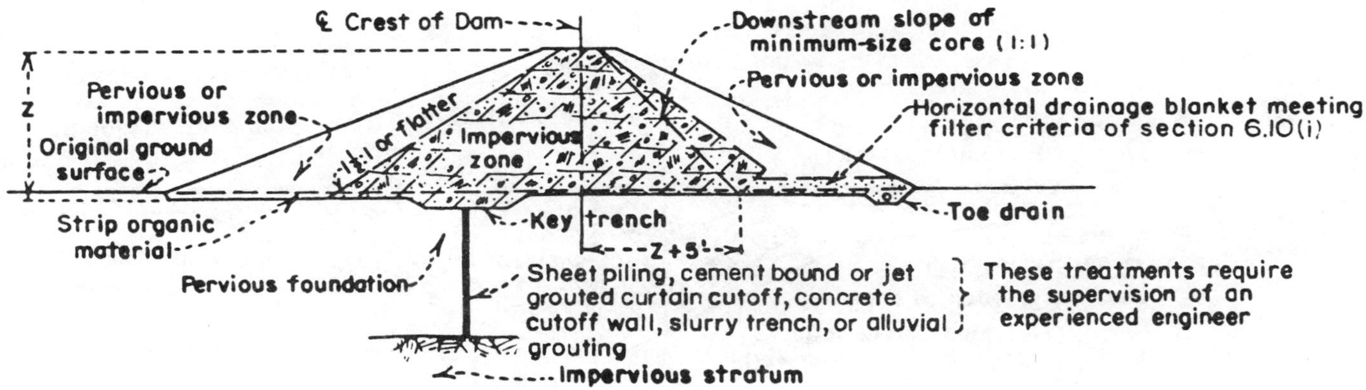

(B) INTERMEDIATE DEPTH OF PERVIOUS FOUNDATION

FIGURE A-32 Deep exposed pervious foundation.

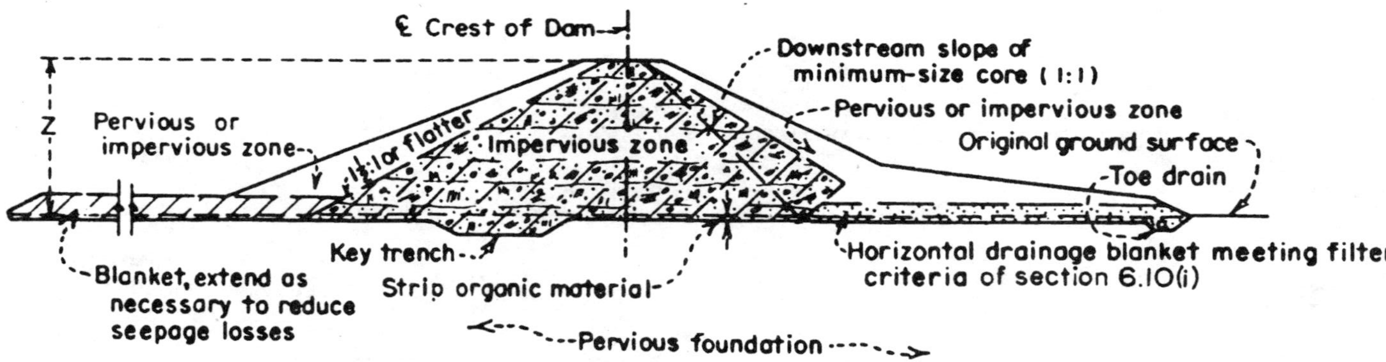

(C) DEEP PERVIOUS FOUNDATION

FIGURE A-33 Deep exposed pervious foundation.

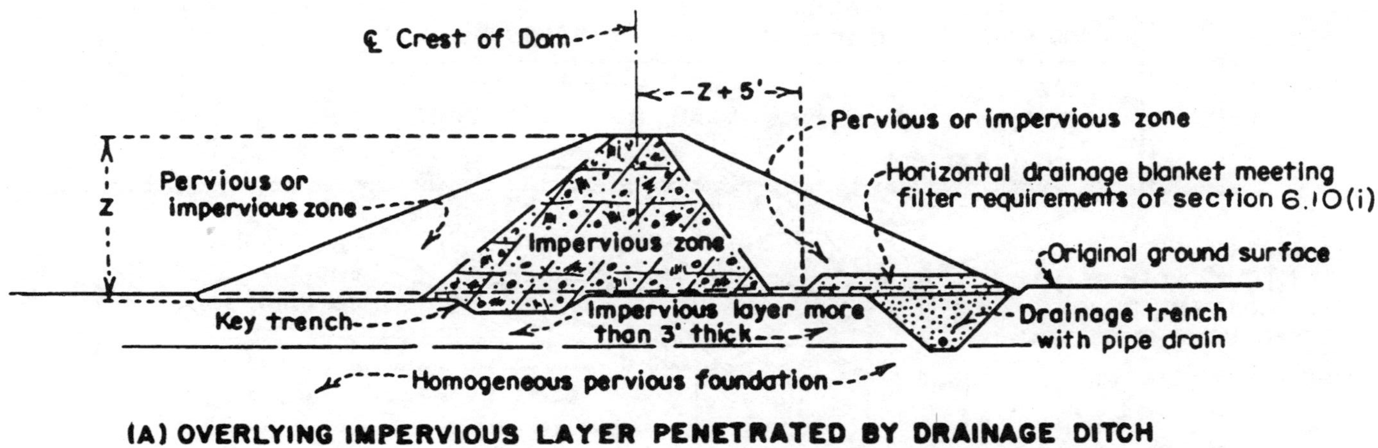

FIGURE A-34 Overlying impervious layer penetrated by drainage ditch.

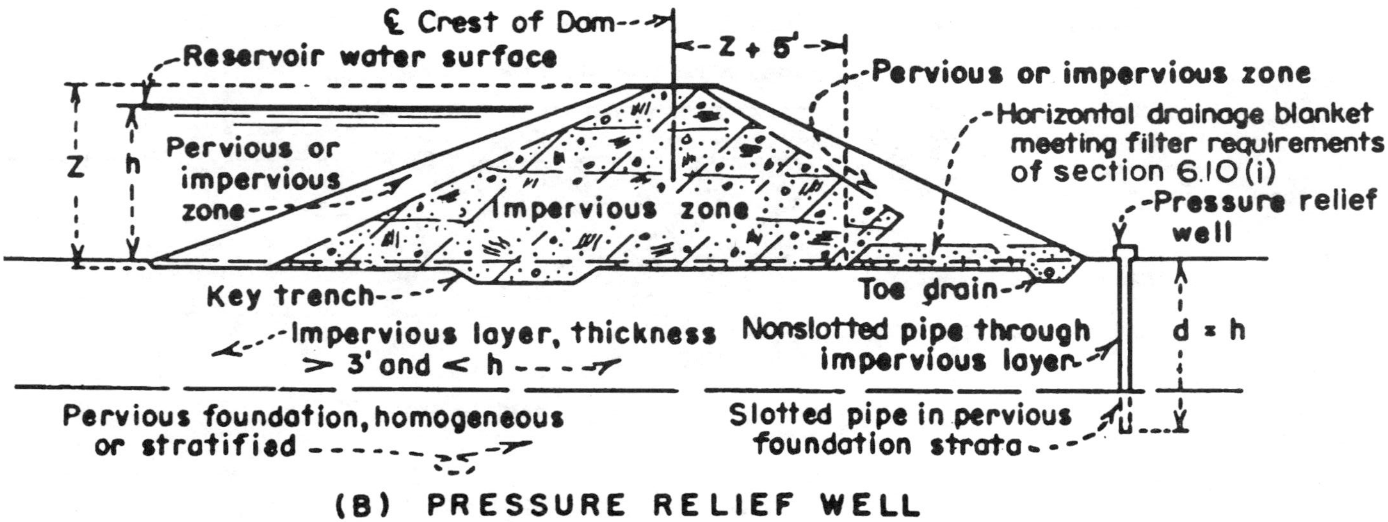

FIGURE A-35 Pressure relief well.

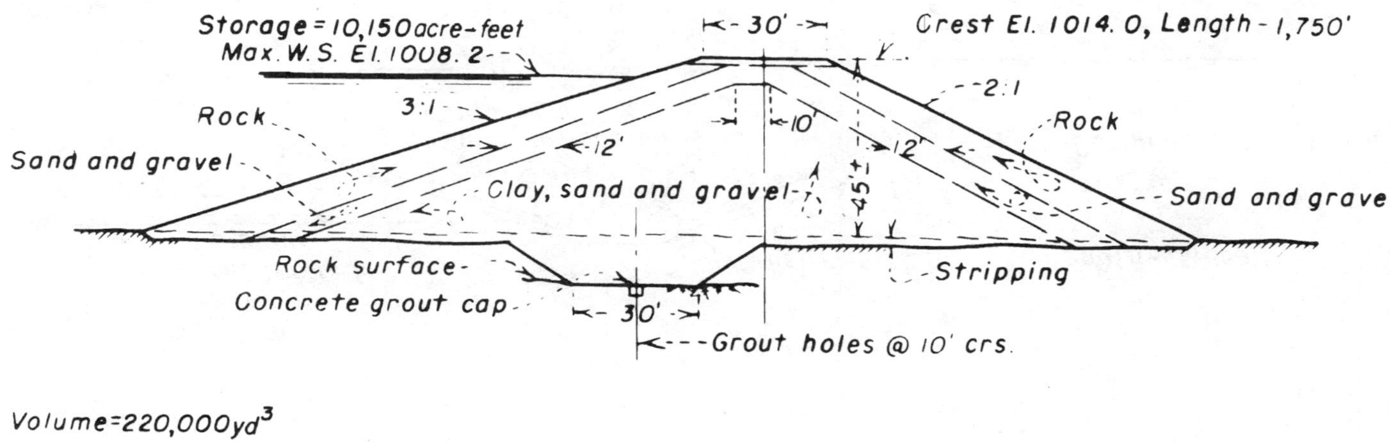

FIGURE A-36a Soda Lake Dike. Located offstream from the Columbia River, Washington.

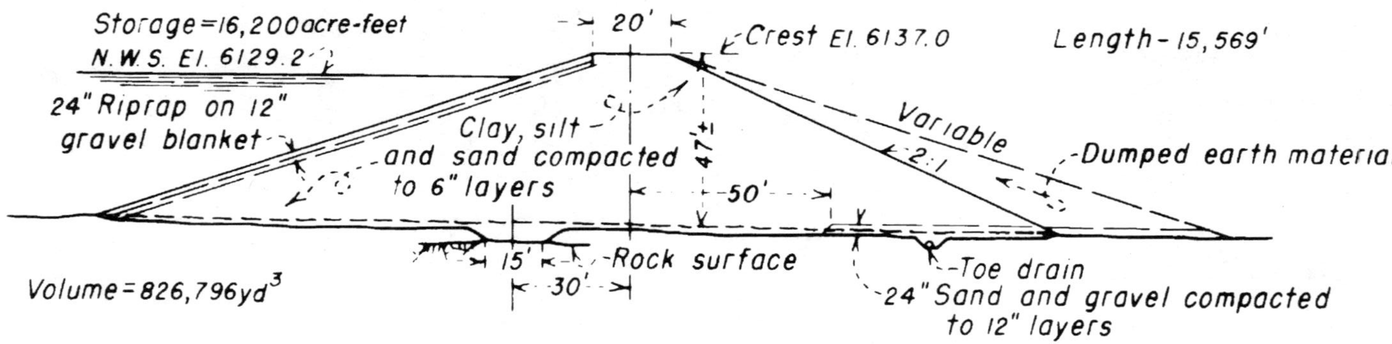

FIGURE A-36b Stubblefield Dam. Located offstream from the Vermejo River, New Mexico.

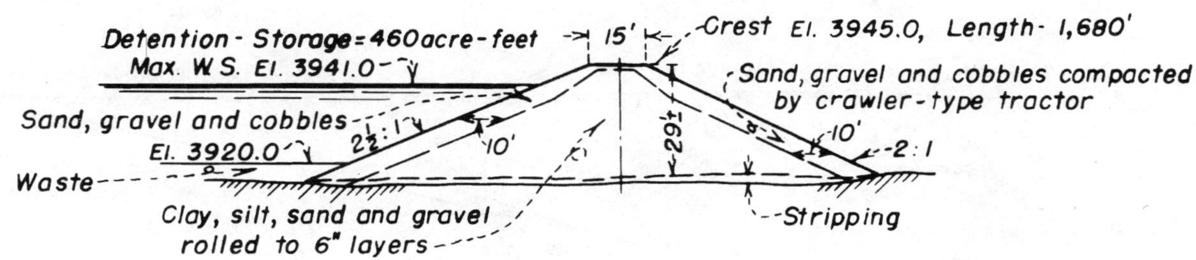

Volume = 85,940 yd³

Picacho South Dam. A detention dam located on the South Branch of Picacho Arroyo, New Mexico (constructed 1953-54). From 103–D–586.

FIGURE A-37 Picacho South Dam. A detention dam located on the south branch of Picacho Arroyo, New Mexico.

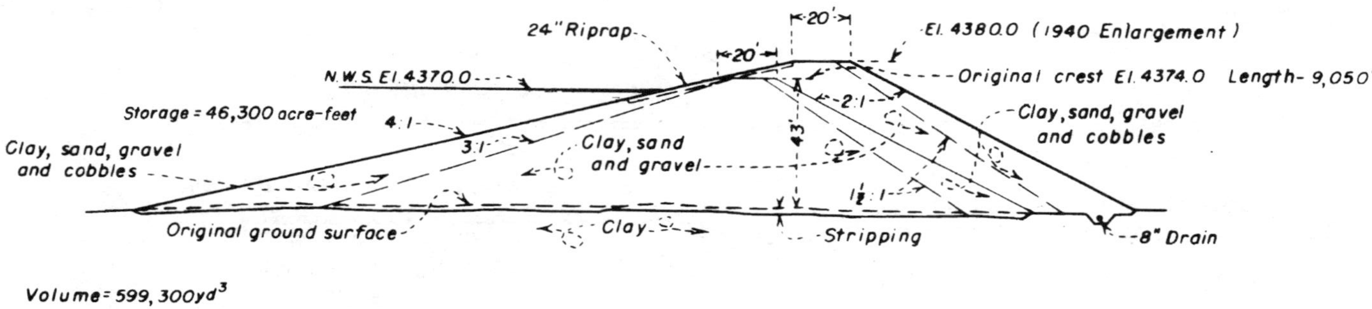

Pishkun Dikes. Located offstream from the Sun River, Montana (constructed 1930-31). From 103–D–580.

FIGURE A-38 Pishkun Dikes. Located offstream from the Sun River, Monatan.

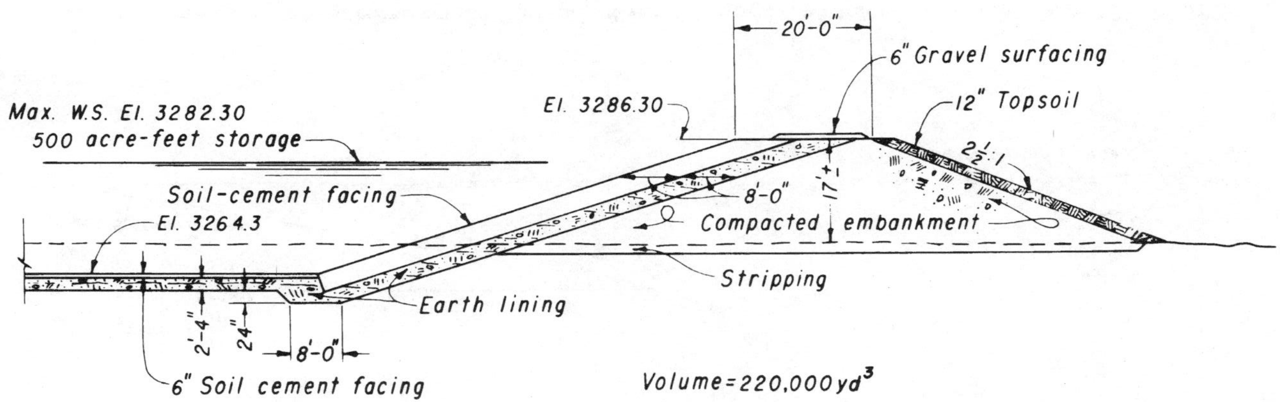

Lubbock Regulating Reservoir. Located offstream near Lubbock, Texas (completed 1966). From 662–D–409.

FIGURE A-39 Lubbock Regulating Reservoir. Located offstream near Lubbock, Texas.

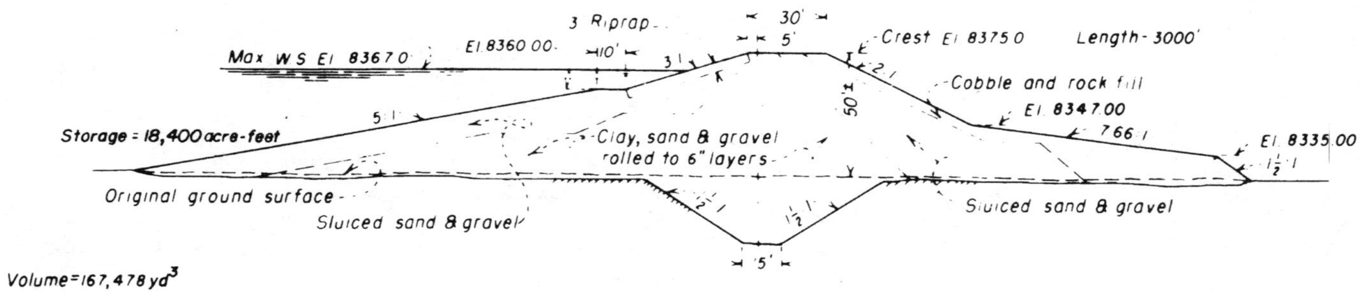

Shadow Mountain Dam. Located on the Colorado River, Colorado (constructed 1943-46). From 103–D–582.

FIGURE A-40 Shadow Moutain Dam. Located on the Colorado River, Colorado.

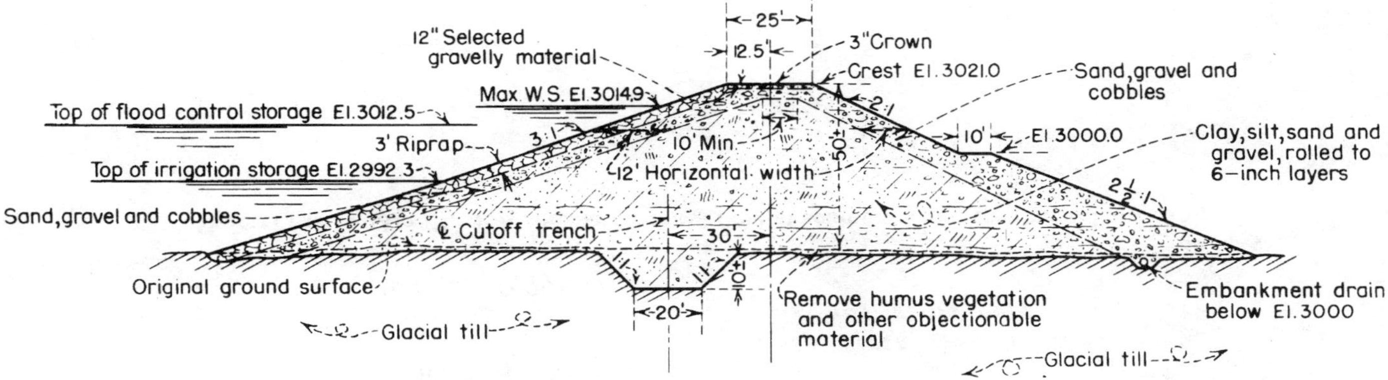

(A) MAXIMUM DIKE SECTION

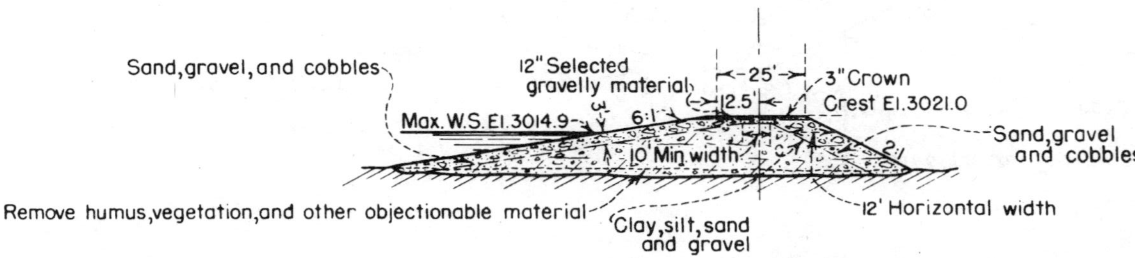

(B) FREEBOARD DIKE SECTION

Sections of Tiber Dike. Located on the Marias River, Montana (constructed 1953-56). 288–D–2501.

FIGURE A-41 Sections of the Tiber Dike. Located on the Marias River, Montana.

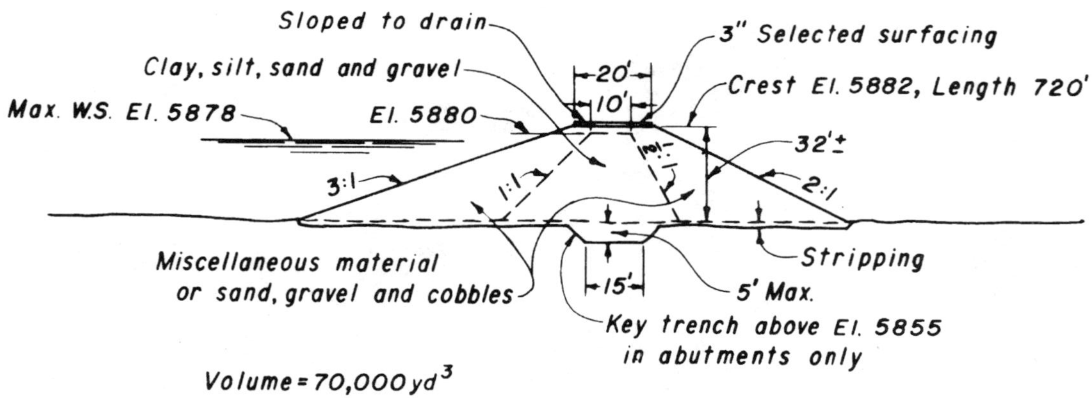

Sheep Creek Barrier Dam. A sediment control dam in southern Utah (constructed in 1956).

FIGURE A-42 Sheep Creek Barrier Dam. a sediment control dam in southern Utah.

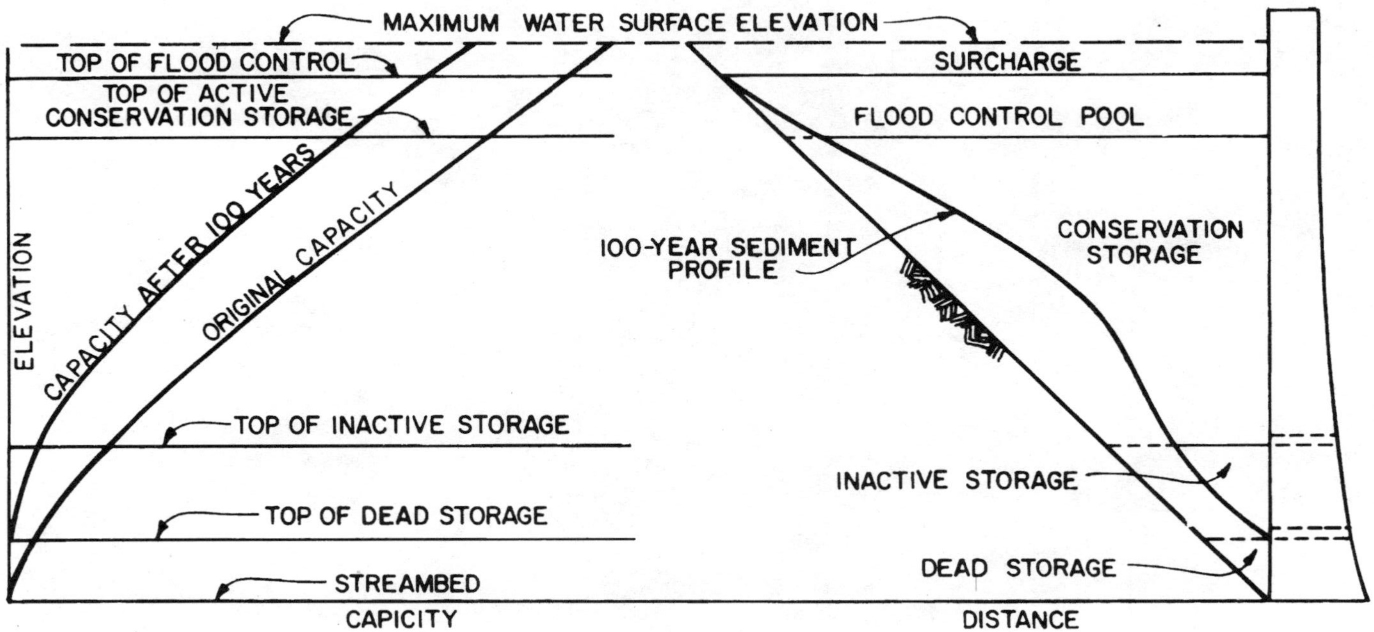

SCHEMATIC DIAGRAM – RESERVOIR ALLOCATIONS, SEDIMENT DEPOSITION

Schematic diagram of reservoir allocations and sediment deposition.

FIGURE A-43 Schematic diagram of reservoir allocations and sediment deposition.

INDEX

A

B

C

D

E

F

G

H

I

M

N

P

R

S

T

U

V

W

NOTES

NOTES

NOTES

NOTES

NOTES

NOTES

NOTES

NOTES

NOTES